Encyclopaedia of Nanotechnology

Vol. 5 : Defence Application

Dr. Rita Khare

AGROTECH PRESS

JAIPUR ▪ NEW DELHI

AGROTECH PRESS

Regd. Off. "Bony Residency", Gate No. 2
Opp. Tilak Public School, Vishweswariya Nagar,
Gopalpura Road, Jaipur - 302018, India
Ph: 0141 - 2761280, Telefax: 2761381
E-mail: agrotechpress@gmail.com

Delhi Office: 102, 1st Floor, Satyam House,
4327/3, Ansari Road, Darya Ganj,
New Delhi - 110002, India
Ph: 011-45652440
visit us: www.agrotechpress.com

Distributed by :
KEYA INTERNATIONAL PRESS LLP
E-mail: keyainternationalpressllp@gmail.com

Published in India, 2024

ISBN : 9789384568566 (10 Vol. Set)

Typeset by Agrotech Press

Printed and bound in India by Replika Press Pvt. Ltd.

Contents

Preface

It is generally agreed that advances in nanotechnology will drive the next paradigm shift in science and technology. Nanotechnology is an area which has highly promising prospects for turning fundamental research into successful innovations. Not only to boost the competitiveness of our industry but also to create new products that will make positive changes in the lives of our citizens, be it in medicine, environment, electronics or any other field. Nanosciences and nanotechnologies open up new avenues of research and lead to new, useful, and sometimes unexpected applications. Novel materials and new-engineered surfaces allow making products that perform better. Whilst many commercial applications of nanotechnology remain theoretical, we now have the capability to manipulate and restructure materials at the nanoscale. Nanotechnology is the understanding and control of matter at dimensions of roughly 1-100 nm, where unique phenomena enable novel applications. Encompassing nanoscale science, engineering and technology, nanotechnology involves imaging, measuring, modelling, and manipulating matter at this length scale. At this level, the physical, chemical and biological properties of materials differ in fundamental and valuable ways from both the properties of individual atoms and molecules or bulk matter. Nanotechnology research is directed toward understanding and creating improved materials, devices and systems that exploit these new properties. These advances are beginning to have repercussions across all areas of technology, and many products containing nanoparticles are already on the market. Industries and governments across the world are investing heavily in nanotechnology.

Nanotechnology holds strong promises for use in the defence industry. Current thinking is that nanotechnology can be used in different ways by soldiers. With the highly promising expectations of nanotechnology for new innovative products, materials and power sources it is evident that nanotechnology can bring many innovations into the defence world. Nanoscience and Nanotechnology are emerging fields of science and technology that are witnessing the emergence of an increasing number of new ideas and applications in defence and military. The important advantages by nanotechnology offered are miniaturisation of existing equipment to allow it to be not only smaller, but lighter, use less energy and be more readily concealable and to develop and adapt new materials for military purposes. The main aims of military research into nanotechnology are to improve medical and casualty care for soldiers, and to produce lightweight, strong and multi-functional materials for use in clothing, both for protection and to provide enhanced connectivity. Military nanotechnology programs cover a wide array of applications to improve the performance of existing weapon systems and to develop new ones.

Currently, nanotechnology is evolving from the basic stage of its development into the applied research stage of technology maturity. This is not to say that real world applications of nanotechnology do not exist. These mainly break down in areas such as coatings, industrial powders, chemicals and carbon nanotubes. Potential military applications and/or uses of nanotechnology could be far more diverse. One has to underline the fact that nanotechnology will not only give rise to genuine nanostructured weapon systems. Rather, nanotechnology is expected to affect various fields of research and disciplines including pharmaceuticals, biology, chemistry, physics, virology, etc. This is the reason why we must be very careful when discussing the expected impacts of nanotechnology in military affairs. Among the first applications of nanotechnology, we can expect to find materials that are lighter and stronger. These will feature different properties than materials available today. In that sense, nanotechnology will first give rise to improvements in existing systems configurations. It must be underlined that such improvements will be of first importance for the defence community. Nanostructured materials will provide the defence industry with the required means to develop lighter, flexible, more agile and more resistant military platforms, including light armored vehicles, tanks, fighter jets, man-transportable micro-unmanned air vehicles (MUAV), etc. Developments in nanotechnology will ensure the manufacture of platforms that will have an ability to adapt to all types of climate conditions and environments. Remarkable mid-term advances are expected to occur in information technologies. Battle systems architectures, lightweight nanonetworks, self-assembled nanosystems will noticeably improve situational awareness capabilities.

Nanotechnology has not yet acquired the status of a sufficiently mature technology to allow analysts to foresee what could be its precise impact on political-military affairs in the coming decades. This focused guide brings together technical expertise from industry, academia, and government to discuss defence technology needs, identify exciting nano advances primed for technology development, forge connections to facilitate and accelerate the development process, identify obstacles to the accelerated transition of nanotechnology and emerging technology innovations, and address the most critical issues in real and emerging needs for defence.

Dr. Rita Khare

1

Molecular Industrialization of Nanotechnology

Molecular industrialized is at the convergence of several powerful technologies. Living cells join molecules together in precise patterns to form functional structures. Rapid prototyping systems form parts directly from blueprints under computer control. Automated factories assemble parts into products. Molecular manufacturing will combine all of these capabilities and more. A technology that could perform precise rapid prototyping and automated assembly on the molecular scale would be able to take advantage of several additional technical benefits. Precise, stiff surfaces can exhibit extremely low friction and wear. Smaller devices not only have increased functional density, but also increased power density and operation frequency-devices with single-nanometre features might have up to a million-fold such advantage over today's machinery. Structures in which most elements are flawless can have significantly higher strength.

A few simple, well-characterized, reliable, constructive operations, repeated many times, can be extremely powerful. That is how computers work-a modern desktop computer can do a trillion transistor operations every day virtually flawlessly. The digital character of covalent bonds implies that fabrication operations can be error-free for all practical purposes. Thus, once a small set of fabrication operations is known, a large range of billion-atom structures can be designed with confidence, just as computer chips can be built out of well-characterized transistors. This is one reason for expecting development to proceed quite rapidly beyond a certain point.

Once a molecular fabricator is built, getting exponential numbers-even kilogram quantities-of nanoscale machines to work in parallel will largely be a matter of mechanical engineering and a few dozen doubling cycles. One embodiment of the manufacturing potential is a nanofactory-a kilogram-scale box containing micro-molar numbers of molecular fabrication systems in regular, easily controlled arrays, producing kilogram-scale products including duplicate nanofactories in a few hours. Proposed nanofactory architectures scale easily to ton-per-hour and beyond. Since fabrication would be totally automated and use non-scarce factories, product complexity would not add to manufacturing cost.

It is worth repeating that the ability to build machines at the nanometre scale conveys massive scaling law advantages, as long as friction and wear can be avoided through superlubricity or biomimesis. Some of the advantages will depend on the material, but if and when nanofabricated diamond-in-vacuum becomes available, the active component of human-scale products will shrink almost to the vanishing point, and even the structure may be reduced by two to three orders of magnitude.

In the absence of any targeted development program for molecular manufacturing, several technologies appear to be moving toward commercially relevant single-atom fabrication with a projected arrival date of four or five decades. However, nanometre-scale covalent-solid manufacturing systems appear to represent a "sweet spot" in technology space, and development programs will likely be targeted at that spot well in advance of its incidental development. A report issued this week by the National Academy of Sciences calls for "defining and focusing on basic experimental steps that are critical to advancing long-term goals" [toward molecular manufacturing] and for funding "experimental demonstrations that link to abstract models and guide long-term vision." This is likely to open the door for more US (and other) federal funding of relevant work.

A New Weapons Race

A nanofactory that could build high-performance products directly from blueprints in a few hours would have many applications. One obvious product family is weapons, including weapons of mass destruction (WMDs). Higher strength, power density, and functional density would improve a number of existing weapon designs. It would also enable new classes of weapons. For example, UAV's in a wide range of sizes could perform surveillance, sabotage, or antipersonnel missions far beyond what is currently contemplated. It appears that "briefcase to orbit" systems will be possible. A small automated airplane (Helios) has been flown up to 96,000 feet. Even a small rocket should be able to attain orbit from that altitude, and a small aircraft should be able to lift it. Initial calculations suggest that the advantages of diamondoid construction would enable a human-portable system, built by a home-appliance scale nanofactory, to put a kilogram of payload into orbit.

The advantages of inexpensive rapid prototyping of complete products should not be underestimated. The development cycle of field-programmable computer chips can be up to two orders of magnitude faster than that of factory-programmed computer chips; product development cycles might speed up similarly. For a number of reasons, a nanofactory-enabled arms race appears less stable than was the US-Soviet nuclear arms race. Rapid development of new types of weapons may outstrip the capacity of strategists to plan for stability. Temporary asymmetries caused by rapid development may tempt first strikes. The wide range of military and paramilitary options may create many pathways for gradual escalation to conflagration. Destruction caused by nanotech-based weapons could be more targeted and contained than nuclear explosions. Likewise, there may be less stigma attached to nano-built weapons. Nanofactory availability is likely to be far more rapid and widespread than nuclear proliferation. Uncertainty about the enemy's capabilities, as well as increased ability to deploy surveillance, may be stabilizing factors, but on balance the outlook for stability seems poor.

All-out war could be extremely destructive. With distributed, easily concealed nanofactories, it would not be easy to destroy an enemy's ability to fight. It would be far easier to destroy things of sentimental value, such as civilian resources-and civilians themselves. Preliminary thought about the number of modes of attack, concealment, and space/time separation between launch and attack makes the task of defending civilians appear hopeless, even if the enemy is far weaker than oneself. A mentality that welcomed martyrdom would appear to be at a distinct advantage against a nation unwilling to commit genocide.

Other Risks

In addition to its perilous impacts on weapons and war, nanotechnology manufacturing also presents issues of concern about surveillance, terrorism, environmental problems, economic upheaval, and more.

With supercomputers and sensors effectively free, worldwide surveillance networks could be created with semi-automated data processing. Unusual events could be flagged for human attention, and objects and people could be tracked through space and time. This may be very tempting to governments as they try to avoid ownership and use of advanced weapons by individuals or small groups. But it would also enable massive governmental oppression.

Should nanofactory-level technology become available to non-government entities, crime and terrorism may become significantly enabled. This could weaken or even destabilize governments and societies.

If nanofactories can build solar energy collectors and feedstock pre-processors, it is not obvious what scarcity factor will prevent waste on a massive, even global scale. For example, profligate consumption of energy could lead to large fractions of the planet being covered in solar panels. Even things that are rare today, such as sonic booms from small aircraft, may become common enough to pose environmental problems. If small products were made in large quantity, they could form quantities of "nano-litter" that could be difficult to collect.

The economic consequences of nanofactory technology are diverse and sizeable. If nanofactory products are as efficient and high-performance as expected, they may rapidly out-compete other forms of manufacturing. Nanofactory manufacture near time and place of use would affect transportation and storage industries. Manufacturing industries, of course, could be wiped out. New industries and lifestyles could create indirect economic effects. Rapid economic change could weaken or disrupt societies.

A concern that has been raised about molecular manufacturing technology is that small, self-contained, self-replicating systems (so-called 'grey goo') might multiply in the wild and consume large fractions of the biosphere. Originally, it was feared that a laboratory accident could be enough to release such a device. However, current proposals for manufacturing systems do not include anything remotely like such a device, even during development phases. It is not yet known how much of a problem could arise from deliberate release of free-range self-replicators.

They would probably be quite difficult to design, especially in a small package. They would have essentially no practical use, even as a weapon. However, it did not take long for the first computer virus to be created and released; hobbyists may be a source of concern once the technology becomes accessible to them. More analysis is needed to determine the eventual dangers posed by free-range self-replicators, with special attention paid to water-borne designs. However, the dangers of unstable arms races leading to devastating war and/or unbreakable oppression appear more pressing.

Dealing with the Dangers

There will be no simple solution to dealing with molecular manufacturing. Even a policy of massive suppression of it and all possible enabling technologies would not be immune to the possibility of internal instability on the part of the administrative group.

It is also not clear that suppression is desirable. First, the benefits of a mature molecular manufacturing technology could be immense. The ability, for example, to perform planet-scale engineering projects in a matter of months may be necessary to mitigate climate change. Rapid prototyping of nanometre-scale products could accelerate medical research in several ways, as well as providing new kinds of treatment. Huge improvements in sustainable agriculture and efficient distribution systems can be anticipated for countries in the developing world.

Replacement of inefficient infrastructure, along with inexpensive solar collectors, could greatly reduce our future ecological footprint (at least until new uses are found for the newly abundant resources). Inexpensive access to space could provide numerous benefits.

More directly, widespread access to molecular manufacturing would enable widespread development of defences against malicious uses. Although in a military context, it appears that offence will likely beat defence, it may be that widely deployed personal-scale defences can mitigate personal-scale attacks. This possibility needs further analysis.

Preventing malicious or irresponsible people from doing intolerably bad things is only one of the problems. Another source of problems is vicious cycles in social or political systems that may result from the stimulus of near-unlimited manufacturing. Perhaps the direst peril is the unstable arms race described above. Another possible vicious cycle is wealth concentration, caused by a large disparity between manufacturing cost and value to consumer, leading to increasing ability of businesses to purchase business-friendly policy. A third vicious cycle may arise from attempts to prevent ownership of desirable technology: a black market infrastructure may develop and grow.

Unfortunately, almost no resources have been spent on studying these issues. If extensive research and analysis will be necessary to implement wise policy, there may not be time to complete them once the changes start. Large, especially international, administrative bodies also take time to create. If molecular manufacturing is physically possible, then it will happen soon, and technical and policy studies are urgently needed in advance of that date.

Conclusion

Although the possibility of high-performance molecular manufacturing has not been conclusively proved, analysis to date indicates that it will not only work, but will be far more powerful than other technologies, including other nanotechnologies. The actual development schedule is uncertain, but may be expected within the next few decades. Once a general-purpose set of digital molecular fabrication operations is developed, and exponential manufacturing is achieved, further progress toward nanofactories and advanced products may be rapid. Molecular manufacturing is likely to be far more powerful and dangerous than other forms of nanotechnology. Some of the dangers appear comparable in scope to nuclear war. Because some of the dangers arise from systemic vicious cycles and others may result from ill-conceived attempts at policy, studies aimed at developing wise approaches to the problems need to be initiated before the issues arise.

Five-Minute Molecular Manufacturing

Molecular manufacturing refers to a revolutionary near-future manufacturing technology. Whereas today's manufacturing uses large and imprecise machines, molecular manufacturing will use molecular machines to build engineered molecular products. The performance, value, and scope of this technology will be revolutionary and disruptive.

What the Technology Means

The root idea of molecular manufacturing is that molecule-scale fabricators can output their own mass of product in a few minutes. Built from precisely positioned and strongly bonded molecules, the products will be precise and strong. Computer control will enable a wide range of products, including more manufacturing systems. Doubling the number of fabricators every hour would scale a single fabricator into a kilogram-scale personal nanofactory (PN) in a few days. The fully automated PN would contain arrays of fabricators and equipment to join their output into large-scale, integrated, heterogeneous, complete products. Molecular manufacturing goes beyond "nanomanufacturing." A recent NCMS publication defines nanomanufacturing as "the controllable, large-scale manipulation of matter at the nanoscale (0.1 to 100 nanometres), to produce identical value-added components and devices."

This definition does not require the use of nanoscale fabrication systems, without which it will be difficult or impossible to produce large-scale, intricate, integrated products. Nanomanufacturing may produce a kilogram of nanotubes or a nanogram of electronic circuits, but it cannot efficiently produce a one-kilogram supercomputer. A nanofactory, containing vast numbers of nanoscale fabricators working in parallel under computer control, would be able to build a supercomputer in an hour. Nanomanufacturing, for the most part, will build components, inputs to further manufacturing processes; molecular manufacturing will build complete products directly from raw materials.

Molecular machines such as ribosomes and Nadrian Seeman's DNA machine can build molecules today. Molecular manufacturing will extend this capability to build higher-performance engineered molecules, including molecular machines.

One way to build engineered molecules with programmable structure is to attach a variety of building blocks to a polymer (protein, DNA, etc.) in a programmed sequence. The polymer then folds into its desired shape. Another way to build structure is to attach molecular fragments to selected locations on a growing molecular shape.

A personal nanofactory will be able to control the shape of its product molecules, as well as the joining of product nanosystems into a larger product. Simply changing the blueprint will change the output product. This means that it will produce a huge range of products, and will be able to prototype new products very rapidly, leading to a software-like explosion of new product designs.

Nanoscale machines will have extremely high performance. We've already said that manufacturing systems will be able to produce their weight in complex products in a few minutes. Similarly, motors and generators will be thousands of times more powerful, and computers will be millions of times more compact. In almost any human-scale product, the volume occupied by active components will be negligible. Nanofactory-built materials will be far stronger than today's metals.

Molecular Assembly

The build process in the dishes stacked these blocks to make the memory film on the SuperChip, but how were the blocks themselves built? The hard part in this molecular-manufacturing business has got to be at the bottom of the whole process, at the stage where molecules are put together to make large, complex parts.

The Silicon Valley Faire offers simulations of this molecular assembly process, and at no extra charge. From the tourguide, you learn that modern assembly processes are complex; that earlier processes-like those used by Nanofabricators, Inc.-used clever-but-obscure engineering tricks; and that the simplest, earliest concepts were never built. Why not begin at the beginning? A short walk takes you to the Museum of Antique Concepts, the first wing of the Museum of Molecular Manufacturing.

A peek inside the first hall shows several people strolling around wearing loosely fitting jumpsuits with attached goggles and gloves, staring at nothing and playing mime with invisible objects. Oh well, why not join the fools' parade? Stepping through the doorway while wearing the suit is entirely different. The goggles show a normal world outside the door and a molecular world inside. Now you, too, can see and feel the exhibit that fills the hall. It's much like the earlier simulated molecular world: it shares the standard settings for size, strength, and speed. Again, atoms seem 40 million times larger, about the size of your fingertips. This simulation is a bit less thorough than the last was-you can feel simulated objects, but only with your gloved hands. Again, everything seems to be made of quivering masses of fused marbles, each an atom.

"Welcome," says the tourguide, "to a 1990 concept for a molecular-manufacturing plant. These exploratory engineering designs were never intended for actual use, yet they demonstrate the basics of molecular manufacturing: making parts, testing them, and assembling them." Machinery fills the hall. Overall, the sight is reminiscent of an automated factory of the 1980s or 1990s. It seems clear enough what must be going on: Big machines stand beside a conveyor belt loaded with half-finished-looking blocks of some material; the machines must do some sort of work on the blocks. Judging by the conveyor belt, the blocks eventually move from one arm to the next until they turn a corner and enter the next hall.

Since nothing is real, the exhibit can't be damaged, so you walk up to a machine and give it a poke. It seems as solid as the wall of the nanocomputer in the previous tour. Suddenly, you notice something odd: no bombarding air molecules and no droplets of water-in fact, no loose molecules anywhere. Every atom seems to be part of a mechanical system, quivering thermal vibration, but otherwise perfectly controlled. Everything here is like the nanocomputer or like the tough little gear; none of it resembles the loosely coiled protein or the roiling mass of the living cell.

The conveyor belt seems motionless. At regular intervals along the belt are blocks of material under construction: workpieces. The nearest block is about a hundred marble-bumps wide, so it must contain something like 100 x 100 x 100 atoms, a full million. This block looks strangely familiar, with its rods, crank, and the rest. It's a nanocomputer-or rather, a block-like part of a nanocomputer still under construction. Standing alongside the pieces of nanocomputer on the conveyor belt, dominating the hall, is a row of huge mechanisms. Their trunks rise from the floor, as thick as old oaks. Even though they bend over, they rear overhead. "Each machine," your tourguide says, "is the arm of a general-purpose molecular assembler.

One assembler arm is bent over with its tip pressed to a block on the conveyor belt. Walking closer, you see molecular assembly in action. The arm ends in a fist-sized knob with a few protruding marbles, like knuckles. Right now, two quivering marbles-atoms-are pressed into a small hollow in the block. As you watch, the two spheres shift, snapping into place in the block with a quick twitch of motion: a chemical reaction. The assembler arm just stands there, nearly motionless. The fist has lost two knuckles, and the block of nanocomputer is two atoms larger. The tourguide holds forth: "This general-purpose assembler concept resembles, in essence, the factory robots of the 1980s. It is a computer-controlled mechanical arm that moves molecular tools according to a series of instructions. Each tool is like a single-shot stapler or rivet gun. It has a handle for the assembler to grab and comes loaded with a little bit of matter-a few atoms-which it attaches to the workpiece by a chemical reaction." This is like the rejoining of the protein chain in the earlier tour.

Molecular Precision

The atoms seemed to jump into place easily enough; can they jump out of place just as easily? By now the assembler arm has crept back from the surface, leaving a small gap, so you can reach in and poke at the newly added atoms. Poking and prying do no good; when you push as hard as you can (with your simulated fingers as strong as steel), the atoms don't budge by a visible amount. Strong molecular bonds hold them in place. Your pocket tourguide-which has been applying the power of a thousand 1990s supercomputers to the task of deciding when to speak up-remarks, "Molecular bonds hold things together. In strong, stable materials atoms are either bonded, or they aren't, with no possibilities in between. Assemblers work by making and breaking bonds, so each step either succeeds perfectly or fails completely. In pre-breakthrough manufacturing, parts were always made and put together with small inaccuracies. These could add up to wreck product quality. At the molecular scale, these problems vanish. Since each step is perfectly precise, little errors can't add up. The process either works, or it doesn't."

But what about those definite, complete failures? Fired by scientific curiosity, you walk to the next assembler, grab the tip, and shake it. Almost nothing happens. When you shove as hard as you can, the tip moves by about one-tenth of an atomic diameter, then springs back. "Thermal vibrations can cause mistakes by causing parts to come together and form bonds in the wrong place," the tourguide remarks. "Thermal vibrations make floppy objects bend further than stiff ones, and so these assembler arms were designed to be thick and stubby to make them very stiff. Error rates can be kept to one in a trillion, and so small products can be perfectly regular and perfectly identical. Large products can be almost perfect, having just a few atoms out of place." This should mean high reliability. Oddly, most of the things you've been seeing outside have looked pretty ordinary-not slick, shiny, and perfect, but rough and homey. They must have been manufactured that way, or made by hand. Slick, shiny things must not impress anyone anymore.

Molecular Robotics

By now, the assembler arm has moved by several atom-widths. Through the translucent sides of the arm you can see that the arm is full of mechanisms: twirling shafts, gears, and large, slowly turning rings that drive the rotation and extension of joints along the trunk. The whole system is a huge, articulated robot arm. The arm is big because the smallest parts are the size of marbles, and the machinery inside that makes it move and bend has many, many parts. Inside, another mechanism is at work: The arm now ends in a hole, and you can see the old, spent molecular tool being retracted through a tube down the middle.

Patience, patience. Within a few minutes, a new tool is on its way back up the tube. Eventually, it reaches the end. Shafts twirl, gears turn, and clamps lock the tool in position. Other shafts twirl, and the arm slowly leans up against the workpiece again at a new site. Finally, with a twitch of motion, more atoms jump across, and the block is again just a little bit bigger. The cycle begins again. This huge arm seems amazingly slow, but the standard simulation settings have shifted speeds by a factor of over 400 million. A few minutes of simulation time correspond to less than a millionth of a second of real time, so this stiff, sluggish arm is completing about a million operations per second.

Peering down at the very base of the assembler arm, you can get a glimpse of yet more assembler-arm machinery underneath the floor: Electric motors spin, and a nanocomputer chugs away, rods pumping furiously. All these rods and gears move quickly, sliding and turning many times for every cycle of the ponderous arm. This seems inefficient; the mechanical vibrations must generate a lot of heat, so the electric motors must draw a lot of power. Having a computer control each arm is a lot more awkward now than it was in pre-breakthrough years. Back then, a robot arm was big and expensive and a computer was a cheap chip; now the computer is bigger than the arm. There must be a better way-but then, this is the Museum of Antique Concepts.

Building-Blocks into Buildings

Where do the blocks go, once the assemblers have finished with them? Following the conveyor belt past a dozen arms, you stroll to the end of the hall, turn the corner, and find yourself on a balcony overlooking a vaster hall beyond. Here, just off the conveyor belt, a block sits in a complex fixture. Its parts are moving, and an enormous arm looms over it like a construction crane. After a moment, the tourguide speaks up and confirms your suspicion: "After manufacturing, each block is tested. Large arms pick up properly made blocks. In this hall, the larger arms assemble almost a thousand blocks of various kinds to make a complete nanocomputer.

The grand hall has its own conveyor belt, bearing a series of partially completed nanocomputers. Arrayed along this grand belt is a row of grand arms, able to swing to and fro, to reach down to lesser conveyor belts, pluck million-atom blocks from testing stations, and plug them into the grand workpieces, the nanocomputers under construction. The belt runs the length of the hall, and at the end, finished nanocomputers turn a corner-to a yet-grander hall beyond?

After gazing at the final-assembly hall for several minutes, you notice that nothing seems to have moved. Mere patience won't do: at the rate the smaller arms moved in the hall behind you, each block must take months to complete, and the grand block-handling arms are taking full advantage of the leisure this provides. Building a computer, start to finish, might take a terribly long time. Perhaps as long as the blink of an eye.

Molecular assemblers build blocks that go to block assemblers. The block assemblers build computers, which go to system assemblers, which build systems, which-at least one path from molecules to large products seems clear enough. If a car were assembled by normal-sized robots from a thousand pieces, each piece having been assembled by smaller robots from a thousand smaller pieces, and so on, down and down, then only ten levels of assembly process would separate cars from molecules. Perhaps, around a few more corners and down a few more ever-larger halls, you would see a post-breakthrough car in the making, with unrecognisable engine parts and comfortable seating being snapped together in a century-long process in a hall so vast that the Pacific Ocean would be a puddle in the corner...

Just ten steps in size; eight, starting with blocks as big as the ones made in the hall behind you. The molecular world seems closer, viewed this way.

Molecular Processing

Stepping back into that hall, you wonder how the process begins. In every cycle of their sluggish motion, each molecular assembler gets a fresh tool through a tube from somewhere beneath the floor, and that somewhere is where the story of molecular precision begins. And so you ask, "Where do the tools come from?", and the tourguide replies, "You might want to take the elevator to your left."

Stepping out of the elevator and into the basement, you see a wide hall full of small conveyor belts and pulleys; a large pipe runs down the middle. A plaque on the wall says, "Mechanochemical processing concept, circa 1990." As usual, all the motions seem rather slow, but in this hall everything that seems designed to move is visibly in motion. The general flow seems to be away from the pipe, through several steps, and then up through the ceiling toward the hall of assemblers above.

After walking over to the pipe, you can see that it is nearly transparent. Inside is a seething chaos of small molecules: the wall of the pipe is the boundary between loose molecules and controlled ones, but the loose molecules are well confined. In this simulation, your fingertips are like small molecules. No matter how hard you push, there's no way to drive your finger through the wall of the pipe. Every few paces along the pipe a fitting juts out, a housing with a mechanically driven rotating thing, exposed to the liquid inside the pipe, but also exposed to a belt running over one of the pulleys, embedded in the housing. It's hard to see exactly what is happening.

The tourguide speaks up, saying, "Pockets on the rotor capture single molecules from the liquid in the pipe. Each rotor pocket has a size and shape that fits just one of the several different kinds of molecule in the liquid, so the process is rather selective. Captured molecules are then pushed into the pockets on the belt that's wrapped over the pulley there, then-"

"Enough," you say. Fine, it singles out molecules and sticks them into this maze of machinery. Presumably, the machines can sort the molecules to make sure the right kinds go to the right places.

The belts loop back and forth carrying big, knobby masses of molecules. Many of the pulleys-rollers?-press two belts together inside a housing with auxiliary rollers. While you are looking at one of these, the tourguide says, "Each knob on a belt is a mechanochemical-processing device. When two knobs on different belts are pressed together in the right way, they are designed to transfer molecular fragments from one to another by means of a mechanically forced chemical reaction. In this way, small molecules are broken down, recombined, and finally joined to molecular tools of the sort used in the assemblers in the hall above. In this device here, the rollers create a pressure equal to the pressure found halfway to the centre of the Earth, speeding a reaction that-"

"Fine, fine," you say. Chemists in the old days managed to make amazingly complex molecules just by mixing different chemicals together in solution in the right order under the right conditions. Here, molecules can certainly be brought together in the right order, and the conditions are much better controlled. It stands to reason that this carefully designed maze of pulleys and belts can do a better job of molecule processing than a test tube full of disorganised liquid ever could. From a liquid, through a sorter, into a mill, and out as tools: this seems to be the story of molecule processing. All the belts are loops, so the machinery just goes around and around, carrying and transforming molecular parts.

Beyond Antiques

This system of belts seems terribly simple and efficient, compared to the ponderous arms driven by frantic computers in the hall above. Why stop with making simple tools? You must have muttered this, because the tourguide speaks up again and says, "The Special-Assembler Exhibit shows another early molecular-manufacturing concept that uses the principles of this molecule-processing system to build large, complex objects. If a system is building only a single product, there is no need to have computers and flexible arms move parts around. It is far more efficient to build a machine in which everything just moves on belts at a constant speed, adding small parts to larger ones and then bringing the larger ones together as you saw at the end of the hall above."

This does seem like a more sensible way to churn out a lot of identical products, but it sounds like just more of the same. Gears like fused marbles, belts like coarse beadwork, drive shafts, pulleys, machines and more machines. In a few places, marbles snap into new patterns to prepare a tool or make a product. Roll, roll, chug, chug, pop, snap, then roll and chug some more.

As you leave the simulation hall, you ask, "Is there anything important I've missed in this molecular manufacturing tour?" The tourguide launches into a list: "Yes-the inner workings of assembler arms, with drive shafts, worm gears, and harmonic drives; the use of Diels-Alder reactions, interfacial free-radial chain reactions, and dative-bond formation to join blocks together in the larger-scale stages of assembly; different kinds of mechanochemical processing for preparing reactive molecular tools; the use of staged-cascade methods in providing feed-molecules of the right kinds with near-perfect reliability; the differences between efficient and inefficient steps in molecular processing; the use of redundancy to ensure reliability in large systems despite sporadic damage; modern methods of building large objects from smaller blocks; modern electronic nanocomputers; modern methods for-"

> *"Enough!" you say, and the tourguide falls silent as you pitch it into a recycling bin. A course in molecular manufacturing isn't what you're looking for right now; the general idea seems clear enough. It's time to take another look at the world on a more normal scale. Houses, roads, buildings, even the landscape looked different out there beyond the Faire dome-less crowded, paved, and plowed than you remember. But why? The history books (well, they're more than just books) say that molecular manufacturing made a big difference; perhaps now the changes will make more sense. Yes, it's time to leave.*
>
> *As you toss your goggled, gloved jumpsuit into another bin, a striking dark-haired woman is taking a fresh one from a rack. She wears a jacket emblazoned with the name "Desert Rose NanoManufacturing."*

"How'd you like it?" she asks with a smile.

"Pretty amazing," you say.

"Yes," she agrees. "I saw this sim back when I was taking my first molecular-manufacturing class. I swore I'd never design anything so clunky! This whole setup really brings back the memories-I can't wait to see if it's as crude as I remember." She steps into the simulation hall and closes the door.

Nanotechnology is the science to engineer next generation materials, smart functional devices and structures at a nanometre scale, which is a billionth of a metre. Nanosized particles can be arranged in a variety of configuration to obtain extremely useful materials properties. Some of the key applications of nanomaterials include; cosmetics, pigments for paints, cellular antennas, brazing alloys, filters and membranes, magnetic devices, biomedical sensors and drug delivery systems, nanoelectronics-(MEMS/flat panels) and optics, nozzles, wear resistant coatings, weapons, battery and fuel cell electrodes, thermal management, automotive, and aerospace components. The economic impact as estimated by the National Science Foundation is over $900 billion for different industries in next 10-15 years.

In an effort to harness the advantages offered by nanosize particles, it is extremely essential to understand the assembly or manufacturing techniques at nanoscale. The technology for Nanomanufacturing can be easily termed as adolescent, if not mature. There is a lack of manpower knowledgeable in nanotechnology and Nanomanufacturing. Knowledge in the Universities is also in its fledgling stages as the new courses are being designed and intensive research is being conducted only in last 5 year. Hence, there is an urgent need to fill the vacuum created due to such lack of knowledge. This one-day short course will provide an overview of the nanotechnology and its societal impact,

the major Nanomanufacturing processes, and focus on the selection of a suitable manufacturing processes and materials for specific design and applications. This course will brings an understanding of engineering the nanomaterials with an application oriented philosophy without going deep into mathematical and analytical complexities of Nanomanufacturing.

Exploring the Molecular World

In a scenario in the last chapter, we saw Joel Gregory manipulating molecules in the virtual reality of a simulated world using video goggles, tactile gloves, and a supercomputer. The early twenty-first century should be able to do even better. Imagine, then, that today you were to take a really long nap, oversleep, and wake up decades later in a nanotechnological world.

In the twenty-first century, even more than in the twentieth, it's easy to make things work without understanding them, but to a newcomer much of the technology seems like magic, which is dissatisfying. After a few days, you want to understand what nanotechnology is, on a gut level. Back in the late twentieth century, most teaching used dry words and simple pictures, but now-for a topic like this-it's easier to explore a simulated world. And so you decide to explore a simulation of the molecular world.

Looking through the brochure, you read many tedious facts about the simulation: how accurate it is in describing sizes, forces, motions, and the like; how similar it is to working tools used by both engineering students and professionals; how you can buy one for your very own home, and so forth. It explains how you can tour the human body, climb a bacterium, etc. For starters, you decide to take an introductory tour: simulations of real twentieth-century objects alongside quaint twentieth-century concepts of nanotechnology.

After paying a small fee and memorizing a few key phrases (any variation of "Get me out of here!" will do the most important job), you pull on a powersuit, pocket a Talking Tourguide, step into the simulation chamber, and strap the video goggles over your eyes. Looking through the goggles, you seem to be in a room with a table you know isn't really there and walls that seem too far away to fit in the simulation chamber.

But trickery with a treadmill floor makes the walk to the walls seem far enough, and when you walk back and thump the table, it feels solid because the powersuit stops your hand sharply at just the right place. You can even feel the texture of the carvings on the table leg, because the suit's gloves press against your fingertips in the right patterns as you move. The simulation isn't perfect, but it's easy to ignore the defects. On the table is (or seems to be) an old 1990s silicon computer chip.

Vision and Motion

You feel as though you're falling toward the chip's surface, shrinking rapidly. The world grows blurrier, then everything seems to go wrong as you approach the molecular level. First, your vision blurs to uselessness-there is light, but it becomes a featureless fog. Your skin is tickled by small impacts, then battered by what feel like hard-thrown marbles. Your arms and legs feel as though they are caught in turbulence, pulling to and fro, harder and harder. The ground hits your feet, you stumble and stick to the ground like a fly on flypaper, battered so hard that it almost hurts. You asked for realism, and only the built-in safety limits in the suit keep the simulated thermal motions of air molecules and of your own arms from beating you senseless.

"Stop!" gives you a rest from the suit's yanking and thumping, and "Standard settings!" makes the world around you become more reasonable. The simulation changes, introducing the standard cheats. Your simulated eyes are now smaller than a light wave, making focus impossible, but the goggles snap your vision into sharpness and show the atoms around you as small spheres. (Real nanomachines are as blind as you were a moment ago, and can't cheat.) You are on the surface of the 1990s computer chip, between a cell and two blocky nanocomputers. Your simulated body is 50 nanometres tall, about 1/40,000,000 your real size, and the smaller nanocomputer is twice your height.

The simulation keeps bombarding you with air molecules, but the standard settings leave out the sensation of being pelted with marbles. A moment ago you were stuck tight to the ground by molecular stickiness, but the standard settings give your muscles the effective strength of steel-at least in simulation-by making everything around you much softer and weaker.

The tourguide says that the only unreal features of the simulation have to do with you-not just your ability to see and to ignore thermal shaking and bombardment, but also your sheer existence at a size too small for anything so complex as a human being. It also explains why you can see things move, something about slowing down everything around you by a factor of 10 for every factor of 10 enlargement, and by another factor to allow for your being made stronger and hence faster. And so, with your greater strength and some adjustments to make your arms, legs, and torso less sticky, you can stand, see, feel, and take stock of the situation.

Molecular Texture

The ground underfoot, like everything around you, is pebbly with atom-sized bumps the size of your fingertips. Objects look like bunches of transparent grapes or fused marbles in a variety of pretty but imaginary colours. The simulation displays a view of atoms and molecules much like those used by chemists in the 1980s, but with a sharper 3-D image and a better way to move them and to feel the forces they exert. Actually, the whole simulation setup is nothing but an improved version of systems built in the late 1980s-the computer is faster, but it is calculating the same things. The video goggles are better and the whole-body powersuit is a major change, but even in the 1980s there were 3-D displays for molecules and crude devices that gave a sense of touching them. The gloves on this suit give the sensation of touching whatever the computer simulates. When you run a fingertip over the side of the smaller nanocomputer, it feels odd, hard to describe. It is as if the surface were magnetic-it pulls on your fingertip if you move close enough. But the result isn't a sharp click of contact, because the surface isn't hard like a magnet, but strangely soft. Touching the surface is like touching a film of fog that grades smoothly into foam rubber, then hard rubber, then steel, all within the thickness of a sheet of corrugated cardboard.

Moving sideways, your fingertip feels no texture, no friction, just smooth bumps more slippery than oil, and a tendency to get pulled into hollows. Pulling free of the surface takes a firm tug. The simulation makes your atom-sized fingertips feel the same forces that an atom would. It is strange how slippery the surface is-and it can't have been lubricated, since even a single oil molecule would be a lump the size of your thumb. This slipperiness makes it obvious how nano-scale bearings can work, how the parts of molecular machines can slide smoothly.

But on top of this, there is a tingling feeling in your fingers, like the sensation of touching a working loudspeaker. When you put your ear against the wall of the nanocomputer, you flinch back: for a moment, you heard a sound like the hiss of a twentieth-century television tuned to a channel with no broadcast, with nothing but snow and static-but loud, painfully loud. All the atoms in the surface are vibrating at high frequencies, too fast to see. This is thermal vibration, and it's obvious why it's also called thermal noise.

Gas and Liquid

Individual molecules still move too quickly to see. So, to add one more cheat to the simulation, you issue the command "Whoa!", and everything around seems to slow down by a factor of ten. On the surface, you now can see thermal vibrations that had been too quick to follow. All around, air molecules become easier to watch. They whiz about as thick as raindrops in a storm, but they are the size of marbles and bounce in all directions. They're also sticky in a magnet-like way, and some are skidding around on the wall of the nanocomputer. When you grab one, it slips away. Most are like two fused spheres, but you spot one that is perfectly round-it is an argon atom, and these are fairly rare. With a firm grip on all sides to keep it from shooting away like a watermelon seed, you pinch it between your steel-strong fingers. It compresses by about 10 percent before the resistance is more than you can overcome. It springs back perfectly and instantly when you relax, then bounces free of your grip. Atoms have an unfamiliar perfection about them, resilient and unchanging, and they surround you in thick swarms.

At the base of the wall is a churning blob that can only be a droplet of water. Scooping up a handful for a closer look yields a swarm of molecules, hundreds, all tumbling and bumbling over one another, but clinging in a coherent mass. As you watch, though, one breaks free of the liquid and flies off into the freer chaos of the surrounding air: the water is evaporating. Some slide up your arm and lodge in the armpit, but eventually skitter away. Getting rid of all the water molecules takes too much scraping, so you command "Clean me!" to dry off.

Too Small and Too Large

Beside you, the smaller nanocomputer is a block twice your height, but it's easy to climb up onto it as the tourguide suggests. Gravity is less important on a small scale: even a fly can defy gravity to walk on a ceiling, and an ant can lift what would be a truck to us. At a simulated size of fifty nanometres, gravity counts for nothing. Materials keep their strength, and are just as hard to bend or break, but the weight of an object becomes negligible. Even without the strength-enhancement that lets you overcome molecular stickiness, you could lift an object with 40 million times your mass-like a person of normal size lifting a box containing a half-dozen fully loaded oil tankers. To simulate this weak gravity, the powersuit cradles your body's weight, making you feel as if you were floating.

This is almost like a vacation in an orbital theme park, walking with stickyboots on walls, ceilings, and whatnot, but with no need for anti-nausea medication. This looks like a cluster of grapes and is about the same size. It even feels a bit like a bunch of grapes, soft and loose. The parts don't fly free like a gas or tumble and wander like a liquid, but they do quiver like gelatin and sometimes flop or twist. It is solid enough, but the folded structure is not as strong as your steel fingers. In the 1990s, people began to build molecular machinery out of proteins, copying biology. It worked, but it's easy to see why they moved on to better materials.

From a simulated pocket, you pull out a simulated magnifying glass and look at the simulated protein. This shows a pair of bonded atoms on the surface at 10 times magnification. The atoms are almost transparent, but even a close look doesn't reveal a nucleus inside, because it's too small to see. It would take 1,000 times magnification to be able to see it, even with the head start of being able to see atoms with your naked eye. How could people ever confuse big, plump atoms with tiny specks like nuclei? Remembering how your steel-strong fingers couldn't press more than a fraction of the way toward the nucleus of an argon atom from the air, it's clear why nuclear fusion is so difficult. In fact, the tourguide said that it would take a real-world projectile over a hundred times faster than a high-powered rifle bullet to penetrate into the atomic core and let two nuclei fuse. Try as you might, there just isn't anything you could find in the molecular world that could reach into the middle of an atom to meddle with its nucleus. You can't touch it and you can't see it, so you stop squinting though the magnifying glass. Nuclei just aren't of much interest in nanotechnology.

Puzzle Chains

Taking the advice of the tourguide, you grab two molecular knobs on the protein and pull. It resists for a moment, but then a loop comes free, letting other loops flop around more, and the whole structure seems to melt into a writhing coil. After a bit of pulling and wrestling, the protein's structure becomes obvious: It is a long chain-longer than you are tall, if you could get it straight-and each segment of the chain has one of several kinds of knobs sticking off to the side. With the multicoloured, glassy-bead portrayal of atoms, the protein chain resembles a flamboyant necklace. This may be decorative, but how does it all go back together? The chain flops and twists and thrashes, and you pull and push and twist, but the original tight, solid packing is lost. There are more ways to go wrong in folding up the chain than there are in solving Rubik's Cube, and now that the folded structure is gone, it isn't even clear what the result should look like. How did those twentieth-century researchers ever solve the notorious "protein folding problem"? It's a matter of record that they started building protein objects in the late 1980s.

This protein molecule won't go back together, so you try to break it. A firm grip and a powerful yank straightens a section a bit, but the chain holds together and snaps back. Though unfolding it was easy, even muscles with the strength of steel-the strength of Superman-can't break the chain itself. Chemical bonds are amazingly strong, so it's time to cheat again. When you say, "Flimsy world-one second!" while pulling, your hands easily move apart, splitting the chain in two before its strength returns to normal.

You've forced a chemical change, but there must be easier ways since chemists do their work without tiny superhands. While you compare the broken ends, they thrash around and bump together. The third time this happens, the chain rejoins, as strong as before. This is like having snap-together parts, but the snaps are far stronger than welded steel. Modern assembler chemistry usually uses other approaches, but seeing this happen makes the idea of molecular assembly more understandable: Put the right pieces together in the right positions, and they snap together to make a bigger structure.

Remembering the "Whoa!" command, you decide to go back to the properly scaled speed for your size and strength. Saying "Standard settings!," you see the thrashing of the protein chain speed up to hard-to-follow blur.

Nanomachines

Nanomachines are devices built from individual atoms. Some researchers believe that nanomachines will one day be able to enter living cells to fight disease. They also hope to one day build nanomachines that will be able to rearrange atoms in order to construct new objects. If they succeed, nanomachines could be used to literally turn dirt into food and perhaps eliminate poverty.

At your feet is a ribbed, ringed cylindrical object about the size of a soup can-not a messy, loosely folded strand like the protein (before it fell apart), but a solid piece of modern nanotechnology. It's a gear like. Picking it up, you can immediately feel how different it is from a protein. In the gear, everything is held in place by bonds as strong as those that strung together the beads of the protein chain. It can't unfold, and you'd have to cheat again to break its perfect symmetry. Like those in the wall of the nanocomputer, its solidly attached atoms vibrate only slightly. There's another gear nearby, so you fit them together and make the atomic teeth mesh, with bumps on one fitting into hollows on the other. They stick together, and the soft, slick atomic surfaces let them roll smoothly.

Underfoot is the nanocomputer itself, a huge mechanism built in the same rigid style. Climbing down from it, you can see through the transparent layers of the wall to watch the inner works. An electric motor an arm-span wide spins inside, turning a crank that drives a set of oscillating rods, which in turn drive smaller rods. This doesn't look like a computer; it looks more like an engineer's fantasy from the nineteenth century. But then, it is an antique design-the tourguide said that the original proposal was a piece of exploratory engineering dating from the mid-1980s, a mechanical design that was superseded by improved electronic designs before anyone had the tools to build even a prototype. This simulation is based on a version built by a hobbyist many years later. The mechanical nanocomputer may be crude, but it does work, and it's a lot smaller and more efficient than the electronic computers of the early 1990s. It's even somewhat faster. The rods slide back and forth in a blur of motion, blocking and unblocking each other in changing patterns, weaving patterns of logic. This nanocomputer is a stripped-down model with almost no memory, useless by itself. Looking beyond it, you see the other block-the one on the left which contains a machine powerful enough to compete with most computers built in 1990.

This computer is a millionth of a metre on a side, but from where you stand, it looks like a blocky building looming over ten stories tall. The tourguide says that it contains over 100 billion atoms and stores as much data as a room full of books. You can see some of the storage system inside: row upon row of racks containing spools of molecular tape somewhat like the protein chain, but with simple bumps representing the 1s and 0s of computer data.

These nanocomputers seem big and crude, but the ground you're now standing on is also a computer-a single chip from 1990, roughly as powerful as the smaller, stripped-down nanocomputer at your side. As you gaze out over the chip, you get a better sense for just how crude things were a few decades ago. At your feet, on the smallest scale, the chip is an irregular mess. Although the wall of the nanocomputer is pebbly with atomic-scale bumps, the bumps are as regular as tile. The chip's surface, though, is a jumble of lumps and mounds. This pattern spreads for dozens of paces in all directions, ending in an irregular cliff marking the edge of a single transistor. Beyond, you can see other ridges and plateaus stretching off to the horizon. These form grand, regular patterns, the circuits of the computer. The horizon-the edge of the chip-is so distant that walking there from the centre would (as the tourguide warns) take days. And these vast pieces of landscaping were considered twentieth-century miracles of miniaturization?

Cells and Bodies

Even back then, research in molecular biology had revealed the existence of smaller, more perfect machines such as the protein molecules in cells. A simulated human cell-put here because earlier visitors wanted to see the size comparisons-its on the chip next to the smaller nanocomputer. The tourguide points out that the simulation cheats a bit at this point, making the cell act as though it were in a watery environment instead of air. The cell dwarfs the nanocomputer, sprawling across the chip surface and rearing into the sky like a small mountain. Walking the nature trail around its edge would lead across many transistor-plateaus and take about an hour. A glance is enough to show

how different it is from a nanocomputer or a gear: it looks organic, it bulges and curves like a blob of liver, but its surface is shaggy with waving molecular chains.

Walking up to its edge, you can see that the membrane wrapping the cell is fluid (cell walls are for stiff things like plants), and the membrane molecules are in constant motion. On an impulse, you thrust your arm through the membrane and poke around inside. You can feel many proteins bumping and tumbling around in the cell's interior fluid, and a crisscrossing network of protein cables and beams. Somewhere inside are the molecular machines that made all these proteins, but such bits of machinery are embedded in a roiling, organic mass. When you pull your arm out, the membrane flows closed behind. The fluid, dynamic structure of the cell is largely self healing. That's what let scientists perform experimental surgery on cells with the old, crude tools of the twentieth century: They didn't need to stitch up the holes they made when they poked around inside.

Even a single human cell is huge and complex. No real thinking being could be as small as you are in the simulation: A simple computer without any memory is twice your height, and the larger nanocomputer, the size of an apartment complex, is no smarter than one of the submoronic computers of 1990. Not even a bendable finger could be as small as your simulated fingers: in the simulation, your fingers are only one atom wide, leaving no room for the slimmest possible tendon, to say nothing of nerves.

For a last look at the organic world, you gaze out past the horizon and see the image of your own, full-sized thumb holding the chip on which you stand. The bulge of your thumb rises ten times higher than Mount Everest. Above, filling the sky, is a face looming like the Earth seen from orbit, gazing down. It is your own face, with cheeks the size of continents. The eyes are motionless. Thinking of the tourguide's data, you remember: the simulation uses the standard mechanical scaling rules, so being 40 million times smaller has made you 40 million times faster. To let you pull free of surfaces, it increased your strength by more than a factor of 100, which increased your speed by more than a factor of 10. So one second in the ordinary world corresponds to over 400 million here in the simulation. It would take years to see that huge face in the sky complete a single eyeblink.

Enough. At the command "Get me out!", the molecular world vanishes, and your feeling of weight returns as the suit goes slack. You strip off the video goggles-and hugely, slowly, blink.

The Silicon Valley Faire

The tour of the molecular world showed some products of molecular manufacturing, but didn't show how they were made. The technologies you remember from the old days have mostly been replaced-but how did this happen? The Silicon Valley Faire is advertised as "An authentic theme park capturing life, work, and play in the early Breakthrough years." Since "work" must include manufacturing, it seems worth a visit.

A broad dome caps the park-"To fully capture the authentic sights, sounds, and smells of the era," the tourguide politely says. Inside, the clothes and hairstyles, the newspaper headlines, the bumper-to-bumper traffic, all look much as they did before your long nap. A light haze obscures the buildings on the far side of the dome, your eyes burn slightly, and the air smells truly authentic. The Nanofabricators, Inc., plant offers the main display of early nanotechnology. As you near the building, the tourguide mentions that this is indeed the original manufacturing plant, given landmark status over twenty years ago, then made the centrepiece of the Silicon Valley Faire ten years later, when... With a few taps, you reset the pocket tourguide to speak up less often.

Basics of Nanotechnology

What is Nanotechnology all About?

Nanotechnology is the engineering of tiny machines-the projected ability to build things from the bottom up inside personal nanofactories (PNs), using techniques and tools being developed today to make complete, highly advanced products. Ultimately, nanotechnology will enable control of matter at the nanometre scale, using mechanochemistry. Shortly after this envisioned molecular machinery is created, it will result in a manufacturing revolution, probably causing severe disruption. It also has serious economic, social, environmental, and military implications. A nanometre is one billionth of a metre, roughly the width of three or four atoms. The average human hair is about 25,000 nanometres wide.

What's a Personal Nanofactory?

It's a proposed new appliance, something that might sit on a countertop in your home. To build a personal nanofactory (PN), you need to start with a working fabricator, a nanoscale device that can combine individual molecules into useful shapes. A fabricator could build a very small nanofactory, which then could build another one twice as big, and so on. Within a period of weeks, you have a tabletop model.

Products made by a PN will be assembled from nanoblocks, which will be fabricated within the nanofactory. Computer aided design (CAD) programs will make it possible to create state-of-the-art products simply by specifying a pattern of predesigned nanoblocks. The question of when we will see a flood of nano-built products boils down to the question of how quickly the first fabricator can be designed and built.

MOVIE TIME: A short film called Productive Nanosystems: from Molecules to Superproducts depicts an animated view of a nanofactory and demonstrates key steps in the sample process that converts basic molecules into a billion-CPU laptop computer. The 4-minute streaming video is online here.

What could Nanofactories Produce?

Lifesaving medical robots or untraceable weapons of mass destruction. Networked computers for everyone in the world or networked cameras so governments can watch our every move. Trillions of dollars of abundance or a vicious scramble to own everything. Rapid invention of wondrous products or weapons development fast enough to destabilize any arms race. How does 'mechanochemistry' work? It's a bit like enzymes (if you know your chemistry): you fix onto a molecule or two, then twist or pull or push in a precise way until a chemical reaction happens right where you want it. This happens in a vacuum, so you don't have water molecules bumping around. It's a lot more controllable that way.

So, if you want to add an atom to a surface, you start with that atom bound to a molecule called a "tool tip" at the end of a mechanical manipulator. You move the atom to the point where you want it to end up. You move the atom next to the surface, and make sure that it has a weaker bond to the tool tip than to the surface.

When you bring them close enough, the bond will transfer. This is ordinary chemistry: an atom moving from one molecule to another when they come close enough to each other, and when the movement is energetically favourable. What's different about mechanochemistry is that the tool tip molecule can be positioned by direct computer control, so you can do this one reaction at a wide variety of sites on the surface. Just a few reactions give you a lot of flexibility in what you make.

Why do some scientists dismiss this stuff as science fiction? The whole concept of advanced nanotechnology-molecular manufacturing (MM)-is so complex and unfamiliar, and so staggering in its implications, that a few scientists, engineers, and other pundits have flatly declared it to be impossible. The debate is further confused by science-fictional hype and media misconceptions.

It should be noted that none of those who dismiss MM are experts in the field. They may work in chemistry, biotechnology, or other nanoscale sciences or technologies, but are not sufficiently familiar with MM theory to critique it meaningfully. Many of the objections, including those of the late Richard Smalley, do not address the actual published proposals for MM. The rest are unfounded and incorrect assertions, contradicted by detailed calculations based on the relevant physical laws.

Is Nanotechnology Bad or Good?

Nanotechnology offers great potential for benefit to humankind, and also brings severe dangers. While it is appropriate to examine carefully the risks and possible toxicity of nanoparticles and other products of nanoscale technology, the greatest hazards are posed by malicious or unwise use of molecular manufacturing. CRN's focus is on designing and promoting mechanisms for safe development and effective administration of MM.

If MM is so dangerous, why not just completely ban all research and development?

Viewed with pessimism, molecular manufacturing could appear far too risky to be allowed to develop to anywhere near its full potential. However, a naive approach to limiting R&D, such as relinquishment, is flawed for at least two reasons. First, it will almost certainly be impossible to prevent the development of MM somewhere in the world. China, Japan, and other Asian nations have thriving nanotechnology programs, and the rapid advance of enabling technologies such as biotechnology, MEMS, and scanning-probe microscopy ensures that R&D efforts will be far easier in the near future than they are today. Second, MM will provide benefits that are simply too good to pass up, including environmental repair; clean, cheap, and efficient manufacturing; medical breakthroughs; immensely powerful computers; and easier access to space.

What about "Grey Goo"?

The dangers of self-replicating nanobots-the so-called grey goo-have been widely discussed, and it is generally perceived that molecular manufacturing is uncomfortably close to grey goo. However, the proposed production system that CRN supports does not involve free-floating assemblers or nanobots, but much larger factories with all the nanoscale machinery fastened down and inert without external control. As far as we know, a self-replicating mechanochemical nanobot is not excluded by the laws of physics, but such a thing would be extremely difficult to design and build even with a full molecular manufacturing capability. Fiction like Michael Crichton's Prey might be good entertainment, but it's not very good science.

How soon will molecular manufacturing be developed? Based on our studies, CRN believes that molecular manufacturing could be successfully developed within the next ten years, and almost certainly will be developed within twenty years. Shouldn't we be working on current problems like poverty, pollution, and stopping terrorism, instead of putting money into these far future technologies?

We should do both! Development and application of molecular manufacturing clearly can have a positive impact on solving many of today's most urgent problems. But it's equally clear than MM can exacerbate many of society's ills. Knowing that it may be developed within the next decade or two (which is not "far future"), makes preparation for MM an urgent priority.

Implications of Nanofactory Technology

As a general-purpose manufacturing technology, molecular manufacturing will be able to build highly advanced products useful in almost every field of endeavour, including medicine, environmental remediation, daily life, housing, humanitarian relief, entertainment, computation, transportation, infrastructure, and war. The ability of a personal nanofactory to build another nanofactory in an hour means that manufacturing capacity will not be scarce; because the manufacturing process will be automated and self-contained, the products will be inexpensive to make.

Molecular manufacturing will create many risks, but military implications are perhaps the most perilous. A nation or organisation that could design and test new high-tech weapons quickly, then inexpensively build an almost unlimited number of them, could become a potent military force virtually overnight. There are several reasons to think that an arms race based on this technology would not be stable. This logic could lead to preemptive strikes and massive oppression to prevent opponents from developing the technology.

Societal, medical, environmental, and economic implications of molecular manufacturing also may be extreme. Non-scarce manufacturing raises the possibility of planet-scale engineering, which may mitigate existing environmental damage but also may cause new kinds of damage. Engineered machines smaller than cells will be potent medical tools. The ability to manufacture most products from a simple organic feedstock may reduce demand for transportation, warehousing, mined minerals, and of course traditional manufacturing. The ability to build anything you can download plans for could bring Internet-like problems to the physical world. The timeline of molecular manufacturing development will depend on several factors. What we can say with certainty is that development will become rapidly less expensive and more straightforward; in a decade or so, a large corporation might be able to develop these capabilities in just a few years.

Molecular Manufacturing and the Developing World

Molecular manufacturing (MM) can solve many of the world's current problems. For example, water shortage is a serious and growing problem. Most water is used for industry and agriculture; both of these requirements would be greatly reduced by products made by molecular manufacturing. Infectious disease is a continuing scourge in many parts of the world. Simple products like pipes, filters, and mosquito nets can greatly reduce this problem. Information and communication are valuable, but lacking in many places. Computers and display devices would become stunningly cheap. Electrical power is still not available in many areas. The efficient, cheap building of light, strong structures, electrical equipment, and power storage devices would allow the use of solar thermal power as a primary and abundant energy source. Environmental degradation is a serious problem worldwide. High-tech products can allow people to live with much less environmental impact. Many areas of the world cannot rapidly bootstrap a 20th century manufacturing infrastructure. Molecular manufacturing technology can be self-contained and clean; a single packing crate or suitcase could contain all equipment required for a village-scale industrial revolution.

Finally, MM will provide cheap and advanced equipment for medical research and health care, making improved medicine widely available. Much social unrest can be traced directly to material poverty, ill health, and ignorance. MM can contribute to great reductions in all of these problems, and in the associated human suffering.

Despite limited literature discussing the global implications of molecular manufacturing (MM), the seeds for certain key debates are starting to be sown. They essentially mirror those presented for current and near-term nanotechnology: for what purposes will the technology be developed and used, by whom will it be created and owned, what is the nature of the risks it will bring, and what kind of impact will it have upon the global economy and developing world?

The realisation of MM's central goals would undoubtedly lead to the most concentrated technological 'tsunami' ever witnessed. The unpredictable nature of such a revolution makes answering the previous questions all the more difficult. However, just as biotechnology is commonly used as a yardstick for nanotechnology evaluations (given the continuity of many social concerns and relationship between existing biotechnology capabilities and potential to develop nanotechnology), we argue here that the best way to set up the MM debate, in terms of its impact on developing countries, may be to look at developments and trends in nanotechnology. Central to such analysis is addressing the ways in which nanotechnology creates new possibilities for developing countries in terms of access to technology, potential benefits, risks, and shifting views of science and technology, as well as the imposing of new demands in terms of infrastructure and approaches to science.

While many of the issues MM faces may be similar to those presently developing with nanotechnology, MM offers a revolution of a starkly different magnitude. However, MM still faces an 'identity crisis' in the developed world, and an 'identity absence' in the developing world. This has been further hampered by authors and academics who, in writing articles and papers concerning nanotechnology's impact upon the developing world, mix the two terms and confuse references to the relevant potential impacts. Coupled with hype surrounding 'grey goo', a poor foundation for international discussions in which to include MM has evolved. There need to be consistent efforts by those writing in the area to distinguish the kind of nanotechnology to which they are referring.

However, while the scope of potential impact differs greatly between nanotechnology and MM, history suggests there will be universal patterns in terms of its distribution. Whether MM is developed to either a limited or a full capacity, it is more than likely the majority of the world will never reap its benefits and perhaps even fewer will know or understand its potential.

The location of MM's initial development will play a key role in the resolution of suggestions that MM will increase global corporate control, eradicate natural resource markets to the detriment of developing countries or reduce global inequalities and allow countries to 'leap-frog' the industrial revolution. Some suggest that industrialised countries are more fertile ground for its development, while others believe China and India are in a more likely position than the United States of America and European Union to be the initial producers.

However, as McCarthy notes, "just possessing the technology does not mean that a given state will be able to use it to full advantage". Countries will require an ability to integrate and respond to MM in order to exploit its potential.

In more recent times, debates have extended to consider MM's initial development by both small and big research groups, within private and public spheres, by structured and unstructured research groups, as well as by 'rogue' and 'peaceful' nations.

Assessing nanotechnology, we notice an increasing 'nano-divide' in terms of national research levels and funding, as much amongst the developing countries themselves as between the developed and the developing world. This highlights the limitations of discussing the developing world as if it was an homogeneous grouping. Emerging technology affects regions, countries, populations and communities in vastly disparate ways. Already, 62 countries are engaging with nanotechnology on a national level. Nineteen of them are classified as 'developing' but none of these are from the group of countries classified as 'least developed'. However, via nanotechnology we may be witness to a new form of inexpensive access to niche R&D markets, as demonstrated by development of national nanotechnology programs in countries such as Costa Rica.

Furthermore, when spending is adjusted for purchasing-power parity, Chinese government nanotechnology funding ranks second, internationally, behind the United States of America. Nonetheless, funding (on an absolute basis) remains heavily focussed towards programs in industrialised countries of the European Union, the United States of America and Japan. It appears that smaller national programs will require clear, specific strategic road maps that target niche research and consider significant international collaboration if they wish to be part of the 'nanotechnology revolution'.

Initial applications in nanotechnology, such as in cosmetics, sporting apparatus and clothing, have exposed early indicators as to the nature of research orientation. Increasing levels of private sector patent concentration within the developed world, promoting 'broad brush' patents, have added to developing country concerns. By the time MM arrives, corporate control in areas such as the life sciences will be so strong and patents and other forms of technological control so all-encompassing, that the ability to replicate a multi-functional theranostic kit cheaply and rapidly may have little or no bearing on its ability to reach those in greatest need.

Some have suggested that 'economic abundance' resulting from the realisation of MM will negate such a trend. MacGillivray disagrees, believing that with MM "the economics of production will change... human nature won't." History also disagrees. While economic theory is based on scarcity, the current world is already one of abundance. We 'solved' the 'production issue' long ago. Given the world's current population, we do not need greater production to feed such a number. Rather, we require more equitable distribution of resources, reductions in consumption and the recognition that true 'equality' (at Western standards of living) is not sustainable. As the 2002 United Nations' Human Development report notes, if the per capita energy consumption of developing countries were to rise to even half of that of the advanced industrial economies, the energy reserves of this finite planet soon would be exhausted.

Despite the orientation of early applications, a number of groups remain hopeful that nanotechnology can be 'appropriate' and used to fulfil the Millennium Development Goals. Similarly, 'nanofactories' have been touted as 'appropriate technology' on the basis of their potential to reduce skilled labour and supporting infrastructure requirements. However, it depends whether a nanofactory would reduce skilled or basic labour. If the latter, then it fails Schumacher's criteria for an appropriate technology.

Furthermore, assessments of appropriateness must include assessments of risk. As with nanotechnology, suggestions have been made that MM could promote world peace while others have cautioned that it may, initially, prove quite dangerous. Consistently, reports are finding that research into nanotechnology risk is inadequate and under-funded. With MM offering less chance to both predict and react to issues of risk, the key must be to focus on global capacity building in areas of risk assessment and risk management.

The problem with developed country discussions of nanotechnology for the developing world is that they often don't progress beyond the identification of so-called 'appropriate' applications. Very few people have even arrived at an acceptance that there are 'appropriate applications', full stop. MM continues to struggle for mainstream acceptance and it is, perhaps, more likely that its fruition will come via a gradual extension of nanoscale capabilities, rather than an immediate jump to MM.

In terms of capacity requirements, nanotechnology is changing both the way science is conducted within research teams and promoted to the public. As we move to a scale unable to be viewed by a standard microscope, and less likely to be comprehended, environmental and human health risk become increasingly important issues with which to engage the public. Early commentators have suggested that nanotechnology will reinforce the lessons from the GM-debate: that public participation is crucial to acceptance but 'engineering consent' is not a sustainable means for widespread acceptance and development.

In my experience with researchers in Thailand, India and Australia, there is a strong recognition that capacity development in nanotechnology must incorporate ethicists and legal experts in terms of analysing national regulatory frameworks, maximising competitive advantage and contributing to public dialogue. Finally, nanotechnology is demanding and creating a more interdisciplinary view of science at the research level. Since MM will most probably enter in a climate shaped by nanotechnology's legacy in these areas, it is important that such considerations extend to MM debates.

All these considerations lead me to make two suggestions.

The first suggestion is that we work towards a truly international nanotechnology conference some time around 2010. Such a conference would ideally occur within the United Nations framework and involve official country representatives. It would be of great benefit to the conference, in terms of productivity, if countries submitted 'white papers' prior to the commencement of the proceedings, outlining national capacities, strategies and proposed niche markets in nanotechnology. Such a meeting could also incorporate discussion on the proposed International Convention for the Evaluation of New Technologies and gauge reception of and progress towards MM. The Global Nanotechnology Network as well as the Asia Nano Forum provide good examples and existing infrastructure for any efforts to develop international cooperation and consensus.

The second suggestion is that, replicating a method such as the United Kingdom's Royal Academy of Engineering 'upstreaming' approach, civil society groups should seek to increase nanotechnology and MM exposure at international 'grassroots' events such as the World Social Forum. Such efforts hopefully would pave the way for greater developing country engagement in academic writing and scholarship pertaining to these areas.

Despite significant differences in the potential capabilities of nanotechnology and MM, nanotechnology may offer the greatest insights and means by which to influence the future impact of MM with respect to the developing world. Initially, a clear distinction must be made between nanotechnology and MM. Current nanotechnology indicators point towards increasing concern for the developing world in terms of barriers to technology access, inadequate research into environmental and human health risks, and significant demands in terms of the capacity to respond to and develop nanotechnology. Any study of MM and its potential impacts upon the developing world will gain from an appreciation of the relevant context and developments surrounding nanotechnology.

Nanomaterials Set for Explosive Growth

While nanoparticles are already used to make strong, lightweight plastics, their emerging applications are even more impressive. Here's how they will further enhance plastics and achieve big growth: How do you make plastics lighter and stronger? In recent years, the answer has been to imbue them with nanomaterials-extremely small structures usually from 1 to 100 nanometres, or billionths of a metre, in size. What's more, nanomaterials are not only helping produce robust, lightweight plastic composites but are also key to products as far-ranging as sunscreens and electrically conductive coatings. And these nanoscale materials are set to claim new applications in electronics, food packaging, industrial processing and other fields.

Watch for nanomaterials to figure in heat-and corrosion-resistant plastics as well as in automotive catalysts. In short, nanomaterials are on the verge of explosive growth, and you can expect plastics, among many other applications, to benefit.

Currently, nanotechnology generates $385 million in business per year in the U.S., a number that will likely swell to $3.5 billion by 2008 and $20 billion by 2013, according to a report by Consulting Resources Corp. (CRC), a Massachusetts-based market research firm. In 2008 and beyond, nanomaterials will enjoy "increasingly significant" markets in biomaterials, catalysts, diagnostics and electronics, the CRC study forecasts.

Small startups are responsible for much of the advanced work in nanomaterials. Many of these firms have developed proprietary processes to produce these diminutive particles in extremely consistent sizes, shapes and purities. Some have partnered with large global chemical and plastics suppliers, including Honeywell, DuPont, BASF, GE, Bayer, Rohm and Haas, DSM and

Mitsubishi. Commercial and developmental nanoparticles can be classified into three groups-metal oxides, nanoclays and carbon nanotubes.

Carbon Nanotubes

Carbon nanotubes are currently being used to fill electrically conductive plastics to disperse static charges in manufacturing equipment for computer disk drives and semiconductor wafers. They also eliminate the need for expensive primer coats on plastic automobile body panels, making these plastic parts conductive so that they can be electrostatically spray painted right away. While carbon nanotubes go head to head with graphite fibres in some of these applications, nanotubes require significantly lower loadings and thus better preserve polymer properties.

Nanotubes are also under development for use as flame retardants in plastics. For example, Massachusetts-based Hyperion Catalysis International is developing its multi-wall nanotubes for this particular application. "There are preliminary scientific reports that carbon nanotubes are good flame retardants in polyethylene, polypropylene and EVA (ethylene vinyl acetate)," says Patrick Collins, marketing manager for Hyperion. Additionally, the company is working to use these nanomaterials as conductive components in plastic antennas for high-speed wireless communications devices, where they could rapidly dissipate charges and allow for more efficient functioning.

Behind multi-wall carbon nanotubes in the development timeline are single-wall carbon nanotubes, which possess extraordinarily high electrical and thermal conductivity, strength, stiffness and toughness. While they are extremely challenging to mass-produce, Texas-based startup Carbon Nanotechnologies Inc. is now fabricating commercial quantities at a pilot plant. Single-wall nanotubes could potentially be used as field emitters in flat panel displays as well as in exceptionally strong industrial plastics, which could withstand high temperatures, severe chemicals, corrosive conditions, tremendous pressures and abrasion.

The Nano Future

Indeed, nanomaterials are on the verge of adding a whole new set of applications to an already impressive list that includes sunscreens, electrically conductive coatings and strong, lightweight plastic composites. With a market that many expect to explode, these nanoparticles will become essential to many more products in the near future. And you can expect plastics to improve drastically as a result.

Technical Basis of Molecular Manufacturing

At its foundation, molecular manufacturing works by doing a few precise fabrication operations, very rapidly, at the molecular level, under computer control. It can thus be viewed as a combination of mechanical engineering and chemistry, with some additional help from rapid prototyping, automated assembly, and related fields of research. Atoms and inter-atomic bonds are completely precise: every atom of a type is identical to every other, and there are only a few types. Barring an identifiable error in fabrication, two molecules manufactured according to the same blueprint will be identical in structure and shape (with transient variations of predictable scale due to thermal noise and other known physical effects). This consistency will allow fully automated fabrication. Computer controlled addition of molecular fragments, creating a few well-characterized bond types in a multitude of selected locations, will enable a vast range of components to be built with extremely high reliability. Building with reliable components, higher levels of structure can retain the same predictability and engineerability.

A fundamental "scaling law" of physics is that small systems operate faster than large systems. Moving at moderate speed over tiny distances, a nanoscale fabrication system could perform many millions of operations per second, creating products of its own mass and complexity in hours or even minutes. Along with faster operation comes higher power density, again proportional to the shrinkage: nanoscale machines might be a million times more compact than today's technology.

Computers would shrink even more profoundly, and non-electronic technologies already analyzed could dissipate enough less power to make the shrinkage feasible. Although individual nanoscale machines would have small capacity, massive arrays could work together; it appears that gram-scale computer and motor systems, and ton-scale manufacturing systems, preserving nanoscale performance levels, can be built without running a foul of scaling laws or other architectural constraints including cooling. Thus, products will be buildable in a wide range of sizes. A complete list of advantages and capabilities of molecularly manufactured products, much less an analysis of the physical basis of the advantages, would be beyond the scope of this paper. But several additional advantages should be noted. Precisely fabricated covalent materials will be much stronger than materials formed by today's imprecise manufacturing processes. Precise, well-designed, covalently structured bearings should suffer neither from wear nor from static friction (stiction). Carbon can be an excellent conductor, an excellent insulator, or a semiconductor, allowing a wide range of electrical and electronic devices to be built in-place by a molecular manufacturing system.

Development of Molecular Manufacturing

Although its capabilities will be far-reaching, the development of molecular manufacturing may require a surprisingly small effort. A finite, and possibly small, number of deposition reactions may suffice to build molecular structures with programmable shape—and therefore, diverse and engineerable function. High-level architectures for integrated kilogram-scale arrays of nanoscale manufacturing systems have already been worked out in some detail. Current-day tools are already able to remove and deposit atoms from selected locations in covalent solids. Engineering of protein and other biopolymers is another pathway to molecularly precise fabrication of engineered nanosystem components. Analysis tools, both physical and theoretical, are developing rapidly.

As a general rule, nanoscale research and development capabilities are advancing in proportion to Moore's Law—even faster in some cases. Conceptual barriers to developing molecular manufacturing systems are also falling rapidly. It seems likely that within a few years, a program to develop a nanofactory will be launched; some observers believe that one or more covert programs may already have been launched. It also seems likely that, within a few years of the first success, the cost of developing an independent capability will have dropped to the point where relatively small groups can tackle the project. Without stringent and widespread restrictions on technology, it most likely will not be possible to prevent the development of multiple molecular manufacturing systems with diverse owners.

Products of Molecular Manufacturing

All exploratory engineering in the field to date has pointed to the same set of conclusions about molecular manufacturing-built products:

1. Manufacturing systems can build more manufacturing systems.
2. Small products can be extremely compact.
3. Human-scale products can be extremely inexpensive and lightweight.
4. Large products can be astonishingly powerful.

If a self-contained manufacturing system can be its own product, then manufacturing systems can be inexpensive—even non-scarce. Product cost can approach the cost of the feedstock and energy required to make it (plus licensing and regulatory overhead). Although molecular manufacturing systems will be extremely portable, most products will not include one—it will be more efficient to manufacture at a dedicated facility with installed feedstock, energy, and cooling supplies.

The feature size of nanosystems will probably be about 1 nanometre (nm), implying a million features in a bacteria-sized object, a billion features per cubic micron, or a trillion features in the volume of a ten-micron human cell. A million features is enough to implement a simple CPU, along with sensors, actuators, power supply, and supporting structure. Thus, the smallest robots may be bacteria-sized, with all the scaling law advantages that implies, and a medical system (or weapon system based thereon) could be able to interact with cells and even sub-cellular structures on an equal footing. As a general rule of thumb, human-scale products may be expected to be 100-1000 times lighter than today's versions. Covalent carbon-based materials such as buckytubes should be at least 100 times stronger than steel, and materials could be used more efficiently with more elegant construction techniques. Active components will shrink even more. (Of course, inconveniently light products could be ballasted with water.)

Large nanofactories could build very large products, from spacecraft to particle accelerators. Large products, like smaller ones, could benefit from stronger materials and from active systems that are quite compact. Nanofactories should scale to at least ton-per-hour production rates for integrated products, though this might require massive cooling capacity depending on the sophistication of the nanofactory design.

Possible Weapons Systems

The smallest systems may not be actual weapons, but computer platforms for sensing and surveillance. Such platforms could be micron-scale. The power requirement of a 1-MIPS computer might be on the order of 10-100 pW; at that rate, a cubic micron of fuel might last for 100-1000 seconds. The computer itself would occupy approximately one cubic micron.

Very small devices could deliver fatal quantities of toxins to unprotected humans. Even the smallest ballistic projectiles (bullets) could contain supercomputers, sensors, and avionics sufficient to guide them to targets with great accuracy. Flying devices could be quite small. It should be noted that small devices could benefit from a process of automated design tuning; milligram-scale devices could be built by the millions, with slight variations in each design, and the best designs used as the basis for the next "generation" of improvements; this could enable, for example, UAV's in the laminar regime to be developed without a full understanding of the relevant physics. The possibility of rapid design is far more general than this, and will be discussed below.

The line between bullets, missiles, aircraft, and spacecraft would blur. With lightweight motors and inexpensive manufacturing, a vehicle could contain a number of different disposable propulsion systems for different flight regimes. A "briefcase to orbit" system appears feasible, though such a small device might have to fly slowly to conserve fuel until it reached the upper atmosphere. It might be feasible to use 1 kg of airframe (largely discarded) and 20 kg of fuel (not counting oxidizer) to place 1 kg into orbit; some of the fuel would be used to gather and liquify oxygen in the upper atmosphere for the rocket portion of its flight. (Engineering studies have not yet been done for such a device, and it might require somewhat more fuel than stated here.)

Advanced construction could produce novel energy-absorbing materials involving high-friction mechanical slippage under high stress via micro-or nano-scale mechanical components. In effect, every molecule would be a shock absorber, and the material could probably absorb mechanical energy until it was destroyed by heat.

New kinds of weapons might be developed more quickly with rapid inexpensive fabrication. Many classes of device will be buildable monolithically. For example, a new type of aircraft or even spacecraft might be tested an order of magnitude more rapidly and inexpensively, reducing the cost of failure and allowing further acceleration in schedule and more aggressive experimentation. Although materials and molecular structures would not encompass today's full range of manufactured substances, they could encompass many of the properties of those substances, especially mechanical and electrical properties. This may enable construction of weapons such as battlefield lasers, rail guns, and even more exotic technologies.

Passive armor certainly could not stop attacks from a rapid series of impacts by precisely targeted projectiles. However, armor could get a lot smarter, detecting incoming attacks and rapidly shifting to interpose material at the right point. There may be a continuum from self-reconfiguring armor, to armor that detaches parts of itself to hurl in the path of incoming attacks, to armor that consists of a detached cloud of semi-independent counterweapons.

A new class of weapon for wide-area destruction is kinetic impact from space. Small impactors would be slowed by the atmosphere, but medium-to-large asteroids, redirected onto a collision course, could destroy many square miles. The attack would be detectable far in advance, but asteroid deflection and destruction technology is not sufficiently advanced at this time to say whether a defender with comparable space capacity could avoid being struck, especially if the asteroid was defended by the attacker. Another class of space impactor is lightweight solar sails accelerated to a respectable fraction of light speed by passage near the sun.

These could require massive amounts of inert shielding to stop; it is not clear whether or not the atmosphere would perform this function adequately. A hypothetical device often associated with molecular manufacturing is a small, uncontrolled, exponentially self-replicating system. However, a self-replicating system would not make a very good weapon. In popular conception, such a system could be built to use a wide range of feedstocks, deriving energy from oxidizing part of the material (or from ambient light), and converting the rest into duplicate systems. In practice, such flexibility would be quite difficult to achieve; however, a system using a few readily available chemicals and bypassing the rest might be able to replicate itself—though even the simplest such system would be extremely difficult to design. Although unrestrained replication of inorganic systems poses a theoretical risk of widespread biosphere destruction through competition for resources—the so-called "grey goo" threat—it seems unlikely that anyone would bother to develop grey goo as a weapon, even a doomsday deterrent. It would be more difficult to guide than a biological weapon. It would be slower than a device designed simply to disrupt the physical structure of its target. And it would be susceptible to detection and cleanup by the defenders.

Tactics

A detailed analysis of attack and defence is impossible at this point. It is not known whether sensor systems will be able to effectively detect and repel an encroachment by small, stealthy robotic systems; it should be noted that the smallest such systems might be smaller than a wavelength of visible light, making detection at a distance problematic. It is unknown whether armor will be able to stop the variety of penetrating objects and forces that could be directed at it. Semi-automated R&D may or may not produce new designs so quickly that the side with the better software will have an overwhelming advantage. The energy cost of construction has only been roughly estimated, and is uncertain within at least two orders of magnitude; active systems, including airframes for nano-built weapons, will probably be cost-effective in any case, but passive or static systems including armor may or may not be worth deploying.

Some things appear relatively certain. Unprotected humans, whether civilian or soldier, will be utterly vulnerable to nano-built weapons. In a scenario of interpenetrating forces, where a widespread physical perimeter cannot be established, humans on both sides can be killed at will unless protected at great expense and inconvenience. Even relatively primitive weapons such as hummingbird-sized flying guns with human target recognition and poisoned bullets could make an area unsurvivable without countermeasures; the weight of each gun platform would be well under one gram. Given the potential for both remote and semi-autonomous operation of advanced robotics and weapons, a force with a developed molecular manufacturing capability should have no need to field soldiers; this implies that battlefield death rates will be low to zero for such forces.

A concern commonly raised in discussions of nanotech weapons is the creation of new diseases. Molecular manufacturing seems likely to reduce the danger of this. Diseases act slowly and spread slowly. A sufficiently capable bio-sensor and diagnostic infrastructure should allow a very effective and responsive quarantine to be implemented. Rapid testing of newly manufactured treatment methods, perhaps combined with metabolism-slowing techniques to allow additional R&D time, could minimize disease even in infected persons

Despite the amazing power and flexibility of molecular manufactured devices, a lesson from World War I should not be forgotten: Dirt makes a surprisingly effective shield. It is possible that a worthwhile defencive tactic would be to embed an item to be protected deeply in earth or water. Without active defences, which would also be hampered by the embedding material, this would be at best a delaying tactic. Information is likely to be a key determiner of military success.

If, as seems likely, unexpected offence with unexpected weapons can overwhelm defence, then rapid detection and analysis of an attacker's weapons will be very important. Information-gathering systems are likely to survive more by stealth than by force, leading to a "spy vs. spy" game. To the extent that this involves destruction of opposing spy-bots, it is similar to the problem of defending against small-scale weapons. Note that except for the very smallest systems, the high functional density of molecularly constructed devices will frequently allow both weapon and sensor technology to be piggybacked on platforms primarily intended for other purposes.

It seems likely that, with the possible exception of a few small, fiercely defended volumes, a robotic battleground would consist of interpenetrated forces rather than defencive lines (or defencive walls). This implies that any non-active matter could be destroyed with little difficulty unless shielded heavily enough to outlast the battle.

Strategy

As implied above, a major strategy is to avoid putting soldiers on the battlefield via the use of autonomous or remotely operated weapons. Unfortunately, this implies that an enemy wanting to damage human resources will have to attack either civilian populations or people in leadership positions. To further darken the picture, civilian populations will be almost impossible to protect from a determined attack without maintaining a near-hermetic seal around their physical location. Since the resource cost of such a shield increases as the shield grows (and the vulnerability and unreliability probably also increase), this implies that civilians under threat will face severe physical restrictions from their own defenders.

A substantial variety of attack mechanisms will be available, including kinetic impact, cutting, sonic shock and pressure, plasma, electromagnetic beam, electromagnetic jamming and EMP, heat, toxic or destructive chemicals, and perhaps more exotic technologies such as particle beam and relativistic projectile. A variety of defencive techniques will be available, including camouflage, small size, physical avoidance of attack, interposing of sacrificial mass, jamming or hacking of enemy sensors and computers, and preemptive strike. Many of these offencive and defencive techniques will be available to devices across a wide range of sizes. As explored above, development of new weapon systems may be quite rapid, especially if automated or semi-automated design is employed.

In addition to the variety of physical modes of attack and defence, the cyber sphere will become an increasingly important and complex battleground, as weapon systems increasingly depend on networking and computer control. It remains to be seen whether a major electronic security breach might destroy one side's military capacity, but with increasing system complexity, such an occurrence cannot be ruled out.

Depending on what is being defended, it may or may not be possible to prepare an efficient defence for all possible modes of attack. If defence is not possible, then the available choices would seem to be either preemptive strike or avoidance of conflict. Defence of civilians, as stated above, is likely to be difficult to impossible. Conflict may be avoided by deterrence only in certain cases—where the opponent has a comparable amount to lose. In asymmetric situations, where opponents may have very different resources and may value them very differently, deterrence may not work at all. Conflict may also be avoided by reducing the sources of tension

Broader Context

Military activity does not take place in isolation. It is frequently motivated by non-military politics, though warlords can fight simply to improve their military position. Molecular manufacturing will be able to revolutionize economic infrastructures, creating material abundance and security that may reduce the desire for war—if it is distributed wisely. It appears that an all-out war between molecular manufacturing powers would be highly destructive of humans and of natural resources; the objects of protection would be destroyed long before the war-fighting ability of the enemy. In contrast, a war between molecular manufacturing and a conventionally armed power would probably be rapid and decisive. The same is true against a nuclear power that was prevented from using its nuclear weapons, either by politics or by anti-missile technologies. Even if nuclear weapons were used, the decentralization allowed by self-contained exponentially manufacturing nanofactories would allow survival, continued prosecution of the war, and rapid post-war rebuilding.

The line between policing and military action is increasingly blurred. Civilians are becoming very effective at attacking soldiers. Meanwhile, soldiers are increasingly expected to treat civilians under occupation as citizens (albeit second-class citizens) rather than enemy. At least in the US, paramilitary organisations (both governmental and commercial) are being deployed in internal civilian settings, such as the use of SWAT teams in some crime situations, and Blackwater in post-Katrina New Orleans.

Many molecular manufactured weapon systems will be useable for policing. Several factors will make the systems desirable for police activity: a wide range of weapon effects and intensities to choose from; less risk to police as telepresence is employed; maintaining parity with increasingly armed criminals; and increased deterrence due to increased information-gathering and surveillance. This means that even without military conflict, a variety of military-type systems will be not only developed, but also deployed and used.

It is tempting to think that the absence of nuclear war after six decades of nuclear weapons implies that we know how to handle insanely destructive weapons. However, a number of factors will make molecular manufacturing arms races less stable than the nuclear arms race—and it should be remembered that on several different occasions, a single fortuitous person or event has prevented a nuclear attack. Nuclear weapons are hard to design, hard to build, require easily monitored testing, do indiscriminate and lasting damage, do not rapidly become obsolete, have almost no peaceful use, and are universally abhorred. Molecular manufactured weapons will be easy to build, will in many cases allow easily concealable testing, will be relatively easy to control and deactivate, and would become obsolete very rapidly; almost every design is dual-use, and peaceful and non-lethal (police) use will be common. Nuclear weapons are easier to stockpile than to use; molecular manufactured weapons will be the opposite.

Interpenetrating arrays of multi-scale complex weapons cannot be stable for long. Sooner or later, and probably sooner, a perceived attack will be answered by an actual attack. Whether this mushrooms out of control into a full-scale conflict will depend on the programming of the weapon systems. As long as it is only inanimate hardware at stake, probing attacks and small-scale accidental attacks may be tolerated.

Given the amount of damage that a hostile power armed with molecular manufacturing products could do to the civilian sector, it seems likely that hostile actors will be tolerated only as a last resort, and even apparently non-hostile but untrustworthy actors will be highly undesirable. As mentioned above, an asymmetry in values may prevent deterrence from working. An asymmetry in force, such as between a molecular manufacturing and a pre-MM power, may tempt a preemptive strike to prevent molecular manufacturing proliferation. Likewise, a substantial but decreasing lead in military capability may lead to a preemptive strike. It is unclear whether in general a well-planned surprise attack would lead to rapid and/or inexpensive victory; this may not become clear until offencive and defencive systems are actually developed.

One stable situation appears to be that in which a single power deploys sufficient sensors and weapons to prevent any other power from developing molecular manufacturing. This would probably require substantial oppression of civilians and crippling of industrial and scientific capacity. The government in power would have near-absolute control, being threatened only by internal factors; near-absolute power, combined with an ongoing need for oppression, would likely lead to disastrous corruption.

Widespread recognition of the dangers of arms race, preemptive strike, and war might inspire widespread desire to avoid such an outcome. This would require an unprecedented degree of trust and accountability, worldwide. Current government paradigms are probably not compatible with allowing foreign powers such intimate access to their secrets; however, in the absence of this degree of openness, spying and hostile inspections will only raise tension and reduce trust. One possible solution is for governments to allow their own citizens to observe them, and then allow the information gained by such distributed and non-combative (and thus presumably more trustworthy) observation to be made available to foreign powers.

2

Dendritic Polymers for Polymer Matrix Composites Applications

Dendritic polymers have been examined for several composites applications. Due to the large number of terminal units, the chemistry of dendritic polymers can be tuned to become a very effective coupling agent (a chemical agent that facilitates the adhesion of two components) for fibre-matrix combinations that have been considered incompatible or troublesome to composites engineers. For example, composite materials fabricated using liquid molding methods and vinyl-ester resins are attractive in a number of military applications.

However, the use of higher performance carbon fibres in these vinyl-ester composites has been hindered by less than optimal bonding between vinyl-ester resin and carbon fibres. The use of dendritic polymers as adhesion promoters can improve the bond between the carbon fibres and the vinyl-ester matrix. Dendritic polymers can be incorporated into vinyl-ester carbon fibre composites as fibre sizings (coatings on individual fibres) applied from aqueous solutions to the fibres directly, or alternatively, via a self-assembly method. Self-assembly is a nanoscale process where molecules are inserted into a system and assemble themselves naturally at a desired location. Thus, when the terminal units of dendritic polymers are configured with the proper functional groups at the nanoscale they can be added to the vinyl-ester resin and will naturally migrate to the fibre-matrix interphase via self-assembly.

Another application of dendritic polymers for polymer matrix composites is in creating low-viscosity toughened resins. Cost-reduction efforts in composites manufacturing have led to an increased use of liquid molding techniques for composite materials, and also created a need for low-viscosity toughened resins. Traditional tougheners, including rubbers and thermoplastic polymers, increase resin viscosity to levels that prohibit their use in many of these low cost liquid molding manufacturing methods. Since dendritic polymers have less of an impact on the viscosity of the resin than the traditional tougheners, their judicial use as tougheners in thermosetting resins results in relatively low-viscosity resins that perform comparably to higher-viscosity rubber-modified systems. During cure, micron-sized phases comprised of dendritic polymer-rich material separate from the bulk resin resulting in a toughening mechanism similar to rubber-modified systems.

Dendritic Polymers for Surface Modification

Aware of the growing interest in materials that exhibit specific physiochemical properties and potential applications, we focused our work on modifying commercial agarose with polyfunctional dendrons capable of molecular recognition through hydrogen bonding.

The functional groups on a dendritic polymer can be modified in such a fashion as to create a macromolecule that exhibits self-assembly. For example, these dendritic polymers can be intentionally designed such that when added to an uncured polymer resin system, they migrate to the surface and orient themselves preferentially. An example of where this is useful is in the surface modification of polymer matrix composites to create a specific set of properties on the exposed surface of the material system without sacrificing the composite's bulk properties. This natural migration and orientation of dendritic polymers can be achieved by making some of the functional groups of the dendritic polymer branches incompatible (insoluble) with the resin matrix. Likewise, other functional groups on the same dendritic polymer would be made compatible (soluble) with the resin matrix.

Proper control of the functional groups creates a dendritic material that remains soluble in the uncured resin, while creating a thermodynamic driving force for the incompatible functional groups to migrate the dendritic polymer to the surface of the resin. After curing, there exists a high concentration of dendritic polymers at the surface. Dendritic materials with very strong surface migration tendencies have been developed by modifying the terminal units with low surface energy, perfluorinated functional groups. The aliphatic functional groups, which are compatible with the bulk matrix, permit the homogeneous mixing of the additive and bulk material, and the entire assembly serves to transport the delivered group to the surface in significantly higher concentration than in the bulk material. A practical example utilising nanofunctional dendritic polymers for surface modification is the creation of a self-assembling catalytic layer on the surface of polymers, which could be used to make thin film catalysts, that could potentially be painted onto other surfaces. For instance, a polyethyleneimine dendritic polymer with perfluorinated functional groups and aliphatic chains containing a catalytic metal centre can be mixed within a polyurethane resin. The catalytic dendrimer migrates to the surface of the polyurethane, and upon curing forms a polymer with a catalytic surface layer. At a cast polyurethane surface, the surface properties are dominated by the dendritic polymer additive and the composition of its functional groups. The modified dendritic polymers, complexed with a catalytic metal centre, serves as a nanodelivery system, bringing the catalyst to the polymer surface. The polymer-metal complex has been observed at the surface of a polyurethane film in a quantity approximately ten times greater than its loading in the bulk. These results suggest that small amounts of additive could be used to provide truly functional materials, with active sites embedded or reacted into the polymer. Such an approach has the potential to be more robust than secondary coating treatments such as spray-on applications.

Dendrimer Enhanced Immunoassays

Rapid and reliable detection of biothreat agents is of paramount importance to a variety of defined operational requirements in the biodefence, counterproliferation, and domestic preparedness programs. Current point detection capabilities require expensive equipment and extensively trained personnel to perform their tasks. While many research teams focus their efforts on a "top-down" approach by shrinking existing bulky detectors using a "lab-on-a-chip" strategy, other teams are effectively using a "bottom-up" strategy whereby nanoscale manipulations improve device performance. Dendrimers are ideal nanoscale building blocks for nanomanipulation. Nanoengineered dendrimer-antibody conjugates can be appropriately designed to possess both antibody (protein that binds with high specificity with an antigen, which is a substance that can initiate an immune response) and dendrimer properties because they are about the same size (i.e. 5-10 nm in diameter).

The antibody maintains its primary antigen recognition capability, while the dendrimer strongly adsorbs on a substrate through dendrimer terminal unit surface interactions. The adsorbed conjugate will tend to self-orient the antibody binding domains, known as Fabs (areas on the antibody protein that bind to the antigen), for greater availability in the immunoassay (a technique or system used to detect and possibly identify biological substances, such as biothreat agents, through a binding event with antibodies).

Receptor antibody materials (bot-Fab) are currently used on gold substrates for the detection of biowarfare agents. Because of weak interaction with the gold particles used for the substrate, the gold particles within these bot-Fab/gold complexes tend to aggregate and form clusters. This can obstruct some of the antibody receptor sites from the surrounding environment, reducing their efficiency in detecting biowarfare agents. Research on combining antibody receptor materials into the structure of a dendrimer molecule has shown promise in reducing this problem. The functional groups terminating the branches of dendrimer molecules can be selected so as to enhance surface interaction with the gold nanoparticles (20-70 nm in size).

Furthermore, the structure of the dendrimer/bot-Fab complex can be designed such that the dendrimer strongly adheres to the gold surface while simultaneously forcing orientation of the bot-Fab materials outward so as to maximize their interaction with the environment. Testing of these dendrimer/bot-Fab complexes showed a marked reduction in the aggregation of the gold particles, improving their interaction with and detection of biowarfare agents. A new immunoassay formulation was created using this technology. Testing of the device showed that at least a 20 fold sensitivity enhancement was achieved with the dendrimeranti bot Fab conjugates vs. the anti bot Fab alone. Similarly, production costs might be reduced by as much as 10-fold.

Other Self-Assembled Nanostructured Polymer Materials

This research focuses on the creation of nanostructured polymer and organic-inorganic hybrid materials in order to achieve unusual mechanical, transport, or electro-optical properties. One of the most promising routes to these materials is through the exploitation of the self-assembling characteristics of carefully chosen block copolymer** templates. In utilising these templates it may be important to have control of the orientation (parallel or perpendicular to a substrate) and long range order of the block copolymer morphologies. Innovative techniques, such as the application of electric field or the modification of substrate surface energies, to control orientation and long range order in block copolymer thin films are being explored. These efforts are being performed both internally and through collaborative research with the Army's Polymer Centre of Excellence at the University of Massachusetts.

Permselective Membranes for Protective Clothing

The Army has a need for materials that exhibit selective permeability and favour water transport (for breathability) while inhibiting transport of larger organic molecules such as chemical agents. One approach to achieving this behaviour is through the use of self-assembled block-copolymers. Through appropriate processing and suitable chemical modification, a tri-block copolymer with tailored chemistry and structure has been developed to create a material with nanoscale water transport channels embedded in a barrier matrix of polyisobutylene.

The parent tri-block copolymer is based on a poly(styrene-isobutylene-styrene) that is chemically modified by sulfonation of the styrene domains (blocks). Using this sulfonation process on the styrene domains of a lamellar structure, hydrophyllic nanoscale channels embedded in the polymer material are formed that preferentially favour water transport, while acting as a barrier to other permeants. Because of this attribute, these polymers are being studied for use in protective clothing, serving as a barrier to hazardous materials, while still allowing moisture to escape. These polymers, therefore, have the potential to be used in protective suits that may breathe better than current designs. Further live agent evaluations are in progress on optimized formulations. Textile design and engineering efforts using the novel membranes as a component in protective clothing are currently underway.

Preventive-Arms-Control Considerations

When judging potential military NT uses under criteria of preventive arms control, several of the more generic areas pose no big problems or are too close to civilian uses to consider limitation (e.g., computers). Of the specifically military applications, very few could have positive effects (e.g., sensors for CBW verification). Most, however, raise serious concerns. Existing arms-control agreements can be undermined: the Biological Weapons Convention by new, genetically selective agents, the Treaty on Conventional Armed Forces in Europe by new weapons outside of the treaty definitions.

The international law of warfare is at risk by autonomous fighting systems not reliably recognising non-combatants or combatants hors de combat. Destabilisation is probable with omnipresent sensor nets, micro robots, and autonomous weapons. Arms races are to be expected in all areas of military NT use, including countermeasures and counter-countermeasures. Proliferation of technologies, knowledge, and complete systems is likely, not only to other countries, but also into civilian society. Humans and society could be affected by micro robots used for eavesdropping and spying, or for terrorist and other criminal attacks. Indirectly, civilian barriers against non-medical body manipulation could be circumvented by implants in soldiers' bodies, to achieve faster reaction, a wider sensory spectrum, or better communication and control.

In summary, there are good reasons to devise and implement preventive limits. Taking into account that the technology leader will not keep a monopoly for long, such limits should be in the enlightened national interest of the USA, too.

Proposals for Precautionary Action

Because of the fundamental nature of NT and the potential of misuse also in civilian society, limits and verification of compliance should encompass both, the military and civilian realm. Several steps of analysis and action are recommended.

- Detailed studies for the various military NT areas, including preventive limits.
- Special studies: feasibility of visionary NT (molecular assemblers, nanorobots in the body), feasibility of NT-enabled micro nuclear weapons.
- Investigation of verification means and methods for specific NT limits.
- Increased transparency of NT R&D.
- Unilateral and co-ordinated restraint in military NT, in particular for offencive uses.
- Co-operation of the NT initiatives of various nations in R&D, arms control, safety rules, and societal implications.

The upcoming technologies such as NT, genetics, robotics, and pervasive computing are so powerful that to contain their risks, far-reaching limits with intensive verification and potent criminal prosecution will be needed, on levels similar to those that have evolved within civilian society. It is difficult to conceive of such a system internationally in a world where security is mainly built on the threat of armed force. Long-term security thus calls for strengthening of law and political institutions on the international level, including international criminal law, reducing the dependence on national military forces.

The European Union is demonstrating the success of this approach. Given its focus on civilian technology and problem-solving, Europe may be in a good position to include the major (future) actors in military NT in discussions about preventive limits.

In February 2002, the Massachusetts Institute of Technology submitted a proposal to the U.S. Army for a new research centre devoted to developing military equipment enhanced with nanotechnology. The Army Research Office had issued broad agency solicitations for such a centre in October 2001, and they enthusiastically selected MIT's proposal from among several candidates, awarding them $50 million to kick start what became dubbed the MIT Institute for Soldier Nanotechnologies (ISN). MIT's proposal outlined areas of nanoscience, polymer chemistry, and molecular engineering that could provide fruitful military applications in the near term, as well as more speculative applications in the future. It also featured the striking image of a mechanically armored woman warrior, standing amidst the monuments of some futuristic cityscape, packing two enormous guns and other assault devices. This image proved appealing enough beyond the proposal to grace the ISN's earliest websites, and it also accompanied several publicity announcements for the institute's inauguration.

As this image disseminated, it wasn't long before several comic book fans recognised similarities to the comic book Radix, created by the fraternal team of Ray and Ben Lai. The ISN illustration appeared to be a composite of Radix's fictional heroine, Valerie Fiores, striking her cover pose from Radix #1 (2001), wearing a helmet featured in many issues of this series and superimposed on a cityscape seemingly translated from elsewhere in the same book (Lai and Lai). The ISN illustration had reassembled these fragments from Radix while adding novel flourishes of colouration, background detail, and overlay material. News of this ISN illustration quickly reached Ray and Ben Lai, who threatened lawsuits, stating that MIT had dislocated the futuristic soldier from its real origin in a comic book to the market of nanotechnology research and military investors: "They're selling this as science fact while we're trying to sell it as science fiction... And people don't even know that we created it in the first place. People might even think we're copying them" (Ray Lai qtd. in Shachtman).

The altered image even implies that soldier nanotechnologies are stepping between fictional and real worlds, for the warrior emerges forth from a more transparent area in the background towards a place of more opaque robustness, as if materializing from an ephemeral comic book directly into the battlefields of real science. Or perhaps it's the other way around. Actually, the direction of travel between fiction and reality is not at all clearly indicated by the ISN drawing. But this is no fault of representation, for the image's social context animates such ambiguity. Indeed, the history of nanoscience has depended upon careful navigation of the fault line between novelty and banality, radical visions and technical immediacy, science fiction and science (Milburn, "Nanotechnology"; Hayles; Hessenbruch; Schummer; McCray). Allegorizing this fault line, the ISN drawing effectively serves as a conceptual bridge between the actual and the possible within the arena of nanotechnology (Lopez).

Lawyers for MIT responded that ISN's sampling from the Lais' image fell under legal standards of "fair use" for academic and educational purposes. Moreover, the misappropriation had not hurt sales of Radix; if anything, the resulting scandal only brought new attention to the comic book. The Lais might claim that MIT's representation of their "fantasy" world as "reality" would malign their status as creators of fictions—a grievance noted by the Lais' lawyers in a "cease-and-desist letter" sent to MIT—but recontextualizing "science fiction" as "science fact," according to MIT's legal team, is fully permitted by copyright sanctions for scientific research (Holden; Shachtman). Nevertheless, the ISN illustration was quickly removed from all MIT websites and publications, and on August 30, 2002, Edwin Thomas, the director of ISN, issued a public apology to the Lais through the MIT News Office:

I am writing to apologize publicly to you both [Ray and Ben Lai].... [I]f I had known it was your work, I would not have used it. MIT strongly supports the rights of creators and greatly regrets using the image without permission or credit. I am very sorry that this occurred; it won't happen again. Here is what happened. As my team and I were putting the finishing touches on the proposal, we decided to include a drawing of what the soldier of the future might look like. It was a last minute decision, and I asked my daughter, a graphic artist, to provide an image. We put the image into our proposal to the Army in late February.

In March, the Army made its announcement and MIT included the image in its news release. I didn't know until after your attorney contacted MIT at the end of April that the image apparently was based on your character. As soon as we heard about that, I had it removed from the ISN web pages, notified all of those involved not to use the image for any purpose, and MIT also ordered the image removed from all MIT web sites. (Thomas)

The ISN used the image without knowledge of its science-fictional origin. MIT spokesman Ken Campbell said, "It was an innocent use... We didn't know it was from anyone else's artwork" (qtd. in Frost). Elsewhere, Thomas had asserted that his daughter's interpretive drawing was simply an effort to transform his words into graphics, without any apparent intent to plagiarize: "She did it [the drawing] in a couple of days, and was just trying to illustrate what I had been describing to her" (Kary).

So here the issue has been laid to rest, with the Lais claiming ownership and origin of the image—authorship, that is, of "science fiction"—and Thomas claiming ownership and origin of the technical "description"—authorship, that is, of "science." The picture and the writing are rendered absolutely separate, the one becoming just a convenient "illustration" of the other. Moreover, we are reassured that the science stands completely on its own, for the written elements of the proposal comprise its real "merits" and its real "substance." MIT attorney Ann Hammersla said, "The proposal was peer-reviewed on its technical merits, and the award was not based on that illustration" (qtd. in Russell). An Army spokesperson confirmed that the grant was completely "based on the substance of the proposal, and no illustration was required" (qtd. in Russell; Shachtman). That is, the proposal had been evaluated on the meritorious substance of its words, not on the superfluous excess of its images.

With both parties now appeased, the creators of science and the creators of science fiction can then continue with their own businesses. Comic books and nanotech go their separate ways, as if this chance encounter had been nothing but a strange twist of fate, an accident, a regrettable mistake, a random meeting of words and images without deeper significance.

And yet, while coming down to a case of "illustrative" pictures versus "substantive" words—or even to a case of the profession of science versus the profession of science fiction—this event also manifests the nonlocal cultural mythologies that frame both military technoscience and comic books, exposing their interdependence. Now, there seems little reason to doubt that Thomas's daughter simply landed on the Radix drawing by serendipity, adapting it in all "innocence" as mere "illustration" for her father's description of soldier nanotechnologies, as nothing more than a visual interpretation of the words by whose merits alone the technical application was supposedly judged—an extraneous supplement, that is, to the self-contained text. But we must nevertheless ask why this image seemed for the daughter to be such an appropriate illustration, such an appropriate simulacrum, of otherwise "external" words. Why did her sampling, or poaching, from preexisting images seem to illustrate the description given her so perfectly that the military and the ISN would disseminate this image widely as "a drawing of what the soldier of the future might look like"?

That the cartoon and the text seem to "fit" each other so perfectly—as if they could have been made for each other—should make us suspicious to claims for either scientific or science-fictional autonomy. For both professions here are deeply inhabited and surrounded by traces of a larger media ecology that powerfully informs and motivates them. After all, when Thomas describes his program of soldier nanotechnologies—a program supposedly only illustrated, as if from the outside, by the comic book image—he puts it in the following words:

Our goal is to help greatly enhance the protection and survival of the infantry soldier using nanoscience and nanotechnology.... This will be achieved by creating, then scaling up to a commercial level, revolutionary materials and devices composed of particles or components so tiny that hundreds could fit on the period at the end of this sentence. The idea is to incorporate these nanomaterials and nanodevices into the future soldier's uniform, and associated equipage like helmets and gloves.... [I]magine the psychological impact upon a foe when encountering squads of seemingly invincible warriors protected by armor and endowed with superhuman capabilities, such as the ability to leap over 20-foot walls.

Thomas depicts the development of soldier nanotechnologies as progression from a present "idea" toward a future "goal." In the process, he locates nanotech within the syntactical structure of printed language, imaginarily packing hundreds of nanoscale "components" and "particles" into "the period at the end of this sentence." It would be possible then to see the ISN as a producer of "substantive" text, engineering its very substance—nanotechnology—within the materiality of writing, as if grammatical sign and technical object were indexical.

"The idea" of the whole project to incorporate nanodevices into military uniforms is explained through text alone, as a continuous inscriptional inventory that contains the present in itself—a conveyor of present particles—while simultaneously generating the scientific future and setting it off at a distance. We are asked to "imagine" the future of soldier nanotechnologies as a culmination of alphabetic writing, a giving-forth or materialization of the technical substance abiding within. "Imagine the psychological impact": imagine the invincible powers enabled by those invisible particles at the hypothetical end of this sentence, those particles that are the "end goal" or the "referent" of this sentence as much as they might appear at the space of its final destination, its conclusive period, its full stop. We are directed to think textually, to visualize nanodevices through the medium of print and the analog unreeling of its content towards a deferred future.

FoEA Nano News January 2007

1. US City of Berkley passes first ever local nano laws, second city considers similar
2. Should nano silver underwear be regulated as a pesticide? Should bacteria-killing phone handsets be recalled? Nano silver debate heats up.
3. FoEA speaks at International Council on Nanotechnology conference in Tokyo on need for precautionary management of nanotechnology
4. ETC Group announce winners of nano-hazard symbol contest, releases report "Extreme Genetic Engineering: An introduction to synthetic biology"
5. New FoEA "Avant Card" on nanotechnology – get your hands on one today!
6. Interest grows in nano food; FoEA workshop on nanotechnology and food production at Melbourne Sustainable Living Festival
7. Study finds flexing increases skin penetration by nanoparticles, workers likely to face greater nano exposure risks
8. China determined to challenge west's lead in nanotechnology
9. Israel and India announce nano military programs

As governments world-wide continue to do very little in the face of what is looming as one of the most challenging public policy issues of the 21 century – how to manage the risks and challenges of nanotechnology-local councils and civil society are taking things into their own hands.

In December last year the US City of Berkley declared the first ever local regulation of nanotechnology, and this month the Cambridge City Council has said it is considering similar measures. And while the US Environmental Protection Agency is tying itself up in knots over how it will regulate products that contain nano silver, increasingly recognised as acutely toxic, the ETC Group has announced the winners of its competition to design a globally recognised symbol for nano-hazards.

Meanwhile, Friends of the Earth, with the generous support of Avant Card, have just released a funky new nanotechnology postcard for you to let the Australian Industry Minister know that you support precaution-based management of nanotechnology and public involvement in decision making.

It is clear that the year ahead will be action-packed, as nanotechnology is receiving increasingly critical public attention world-wide. It is encouraging to see the toxicity risks associated with nanomaterials finally receiving some of the attention they warrant. And who knows, perhaps 2007 will be the year that critical discussion of nanotechnology's broader social, military and equity implications joins the debate about how to manage nanotoxicity!

1. US City of Berkley passes first ever local nano laws, second city considers similar

As evidence grows that nanoscale materials, now found in hundreds of consumer products and countless workplaces, may pose new toxicity hazards, local government is becoming frustrated with the inaction of national governments in regulating risks. This has prompted the world's first local government nano regulations, passed in December in Berkeley, California. The regulations will require businesses to provide an annual report to the city's hazardous waste manager that specifies any nanomaterials they use or produce, what procedures they use for tracking, handling and disposing of them, and what they know about how toxic the nanomaterials might be. And now a second US city, Cambridge, has been reported as considering similar regulations. The Berkeley regulations have been criticised as inadequate by those concerned about the need to comprehensively manage the risks of nanotoxicity. And the business community is worried about the emergence of locally different nano regulations generating extra red tape. FoEA continues to call for a comprehensive national, and international, approach to regulating the risks and challenges posed by nanotechnology as the most sensible way to mange its social, economic and political implications, as well as its toxicity risks.

2. Should nano silver underwear be regulated as a pesticide? Should bacteria-killing phone handsets be recalled? The nano silver debate heats up.

The United States Environmental Protection Agency announced last year that because it was legally required to, it would regulate products that contain nano silver, and make claims of anti-microbial properties, as a pesticide. As reported in the January 3 issue of "Environmental Science and Technology", the number of consumer products that contain nano-ingredients has increased by 70 percent since March 2006, and the most common nanomaterial is silver. Widespread use of household products that release nano silver into sewer systems could lead to bioaccumulation of nano silver in fish and die-offs of aquatic life. Not to mention recent studies suggesting that nano silver is also acutely toxic to mammalian cells.

This has caused all sorts of consternation and confusion– will the EPA regulate products that contain nano silver whose manufacturers withdraw marketing claims of antimicrobial action as pesticides? (Incredibly, no). Will the EPA regulate all products that contain nano silver and make antimicrobial claims, such as underwear, phone hand sets, cling wrap or toothpaste, as pesticides? (Unclear).

Will any government body announce moves to introduce a comprehensive regulatory system to manage the actual risks associated with the use of nanomaterials, and the nano content of products, rather than simply the marketing claims made by manufacturers? (We live in hope!)

3. FoEA speaks at International Council on Nanotechnology conference in Tokyo on need for precautionary management of nanotechnology

The mission of the International Council on Nanotechnology (ICON) is "to develop and communicate information regarding potential environmental and health risks of nanotechnology, thereby fostering risk reduction while maximizing societal benefit". Last year I was invited to speak at the ICON conference in Tokyo.

My presentation made three key points: 1) existing evidence of nanotoxicity requires us to take a precautionary approach to risk management – we should not expect workers, the public and the environment to continue facing unregulated exposure to nanomaterials; 2) we must look beyond the question of toxicity risk management to ask critical questions about purpose, predictability, ownership and responsibility (who bears the risks, who stands to gain?); and 3) given the scale of nanotechnology's predicted impacts, we must involve the public in decision making about its introduction.

4. ETC Group announce winners of nano-hazard symbol contest, releases report "Extreme Genetic Engineering: An introduction to synthetic biology"

At the January World Social Forum in Nairobi, Kenya, the ETC Group announced the winners of its international graphic design competition to design a new nano-hazard symbol – comparable to those we have already for nuclear, toxic or other hazardous substances. The world-wide competition for an instantly recognisable hazard warning for nano attracted 482 unique designs from 24 countries. The ETC Group now plans to submit the entries to international standard setting bodies for consideration.

The ETC Group has also released a comprehensive new report "Extreme Genetic Engineering: An introduction to synthetic biology". The report introduces the controversial area of nanobiotechnology and how it is being used to modify DNA, cells and even living organisms. ETC Group explains synthetic biology's implications for bio-weaponry, bio-fuels, intellectual monopoly, trade and the conservation of genetic resources.

5. New FoEA "Avant Card" on nanotechnology – get your hands on one today!

With the generous support of the folk from Avant Card, FoEA's new nano postcard is coming to a café, theatre or other Avant Card outlet near you! The postcard has a colourful, funky "Size Does Matter" on one side, and information about nanotechnology, with a call to the Federal Industry Minister Ian Macfarlane for a moratorium on nanoproducts, and genuine public involvement in nanotechnology decision making on the other. Perfect for passing on to friends who've never heard of nanotechnology, or who have an emerging nano interest. Put one on your fridge, take some into your workplace – and don't forget to send one to the Industry Minister, letting him know we're serious about the need for action on nano!

6. Interest grows in nano food; FoEA workshop on nanotechnology and food production at Melbourne Sustainable Living Festival

In the past couple of months nanotechnology's move from the laboratory to our dinner plates has attracted increasing attention. While the big food companies are refusing to answer questions about their use of nanotechnology, speculation is mounting over which products now contain what nano-ingredients, and importantly, what food regulators are going to do about it. The stakes are high – the food industry anticipates that nanotechnology applications in agriculture, food processing and packaging could prove extremely lucrative – if only they can convince the public to embrace nano foods rather than rejecting them as happened with genetic engineering. However, a recent German consensus conference on nanotechnology has found that not only do the public want "comprehensible labelling, clear definitions, terms and standards as well as far more research into the potential risks before nanotechnology is used to a greater degree in consumer products", they were also "especially critical of the use of nanomaterials in foods."

7. Study finds flexing increases skin penetration by nanoparticles, workers likely to face greater nano exposure risks

US researchers from Rice University and North Carolina State University have published research indicating that repetitive flexing of the skin increases the rate and depth of skin penetration by carbon fullerenes (buckyballs). This research follows an earlier study showing that skin flexing increased skin uptake of beryllium. If skin uptake of nanomaterials occurs, it is possible for nanomaterials to be absorbed into the bloodstream, from where they could access cells, tissues and vital organs. While industry continues to deny that skin uptake of nanomaterials is possible, evidence is growing that skin uptake is clearly possible – and likely – in at least some circumstances. This finding has particular significance for workers who face routine occupational exposure to nanomaterials. Workers who are exposed to nanomaterials while undertaking activities that involve repetitive skin flexing are more likely to experience uptake into the body.

8. China determined to challenge West's lead in nanotechnology

The acknowledged big spenders in the field of nanotechnology are the United States, Europe and Japan. But it is little recognised that even as far back as 2000, Chinese scientists were second only after the US in the number of publications on nanotechnology. A recent report published by the Asian Technology Information Program, based on interviews with leading Chinese nanotechnologists, gives some insights into China's positioning on nanotechnology, its priorities and capacity. The report suggests that Chinese government policy makers and senior scientists believe that nanotechnology is essential for them to "bridge the quality of life "gap" between people living in the rural and urban parts of China. The increasing gap in both the income and the expectations of rural and urban inhabitants has become a major issue for government policy makers."

The report also quotes a "senior nanoscientist, who travels to China frequently" as saying: "If the number of publications is an indicator then, yes, China can compete with the West on basic research. However, it will not have an impact on the West for the next five years at least. The genuine breakthroughs will take time, but my visits to China lead me to conclude that, given the availability of the new hardware and wealth of manpower being injected into the Chinese effort, it is a matter of when and not if, the Chinese will produce commercially useful molecular nano-robots or report on a discovery equivalent to that of carbon nanotubes. The main deficiencies in the Chinese system are technology transfer from the lab to industry and patent rights."

9. Israel and India announce nano military programs

Late last year, Israeli Prime Minister Ehud Olmert was reported as giving the go ahead for a special office to develop a nanotechnology arsenal. The Sydney Morning Herald reports on plans for an "intelligence wasp" or mini drone "that can squeeze into narrow alleys, jam communications, photograph intelligence targets and even kill militants". Additional projects include nano sensors for ambient surveillance and "anti-suicide bomber sensors that can be installed in public places and apparently able to spot a bomber, based on scent, heat and weight". The Indian Express has also reported its government's plans to make use of the Agharkar Research Institute's (ARI) expertise in nanotechnology for military purposes. The Indian Defence Research & Development Organisation (DRDO) has also undertaken initiatives for making the country self-sufficient in a number of strategic materials, e.g. technology for fullerenes and carbon nanotubes which have potential applications in stealth, smart materials and nano-electronics.

Lithography is the process of imprinting patterns on semiconductor materials used in integrated circuits. A pattern, or mask, unique for each circuit, is required. Masks are costly, and new masks are needed each time the circuit design has to be modified. In low-volume situations, mask fabrication can become a significant manufacturing cost. Maskless lithography process using MEMS-based SLMs allows manufactures to build circuits without the expense of individual masks. Lucent's unique SLMs enable such circuits to achieve smaller critical dimensions with higher throughput.

The Bell Labs-developed MEMS SLM technology is made possible by advanced nanofabrication. The Lucent SLMs will contain10 times more individual movable micromirrors, or pixels, than currently available. Individual pixels will be five times smaller and 10 times faster. That means optical maskless lithography systems used in the fabrication of next generation microelectronics can have features as small as 50nm and high throughput.

A MEMS SLM approach makes it possible to manipulate light in ways not previously possible in the fabrication of microdevices. In Lucent's MEMS SLM containing 100 to 200 nanometre features, extremely small mirrors are packed in an integrated multi-megapixel array that reduces requirement for the projection optics needed to achieve small, 50nm, critical dimensions of the next generation integrated circuits.

The lower demagnification required for the smaller mirrors allows for a larger image area with the same optical element sizes and projector numerical aperture. This enables throughput that is 10 to 50-times faster than using other current maskless lithography processes, resulting in efficient and cost-effective fabrication.

"One of the most critical barriers in the areas of maskless lithography is the development and implementation of the appropriate beam modulation technology," said Dave Bishop, vice president of nanotechnology research and president, NJ Nanotech Consortium at Bell Labs. "The micromirror technology Lucent is providing to DARPA is a giant leap forward in the area of advanced MEMS designed and engineering, and is being achieved by combining innovative design with our unique MEMS fabrication capability. We foresee this technology having positive impact on such critical areas as homeland security and military transformational communications.

Regular readers of the *AMPTIAC Quarterly* will have noticed that we've published several 'special issues' over the past few years. The common aim of these publications has been to highlight topics of special interest to targeted technological communities. Examples include our issues on nanotechnology (May 2002) and blast mitigation (*Protecting People at Risk*, February 2003). Both issues were well received by our readers but for different reasons: Nanotechnology represents an exciting and unexplored frontier with intense scientific interest; while protecting people and structures from explosions has gained national attention, especially since 9/11.

This current issue addresses perhaps an even more important topic: the development of new technologies to enable our ground troops to become more effective in the war against terror as well as other emerging global threats.

All one has to do is follow the news reports to appreciate the major technological hurdles now facing the Army. Gone are the days when we faced large standing armies, consisting of heavily armored units employing traditional tactics much like our own. Today, we face adversaries that seemingly have no qualms at sacrificing their lives or those of innocent bystanders in an attempt to inflict damage on our troops. Through a mix of conventional and unconventional weapons (such as improvised explosive devices or IEDs) these fanatics have forced us to adopt new tactics while relying upon our existing weaponry and equipment.

To be totally effective against our new and other possible future enemies, the Army must transform from a force relying on heavy armor to one employing a broad spectrum of lightweight, yet survivable systems and equipment that will enhance their ability to fight. In this context, the word 'transform' means to change doctrine, tactics, and assets to respond rapidly to the environments of the new battlefield. The challenge for our community is to develop the advanced materials that will provide the Army improved effectiveness across the full spectrum of operational environments. To make things even more complex, researchers must address additional 21st Century requirements beyond mere system performance.

They must give greater consideration for 'green' solutions that reduce the generation or introduction of toxic materials into the environment during production, training, deployment, or other military operations. Much recent work has been undertaken to improve the survivability of vehicles and their occupants subjected to fire from ballistic weapons as well as blast and fragmentation from mines, Rocket-Propelled Grenades (RPGs), and IEDs. Discussed in this publication are several of the emerging materials that will enable improved yet lighter armor for future systems. Included are ceramic, metal, and composite material research programs that show tremendous promise. Past armor research has yielded the effective but heavy systems we employ today. Becoming more effective against insurgents requires lighter armored vehicles employing innovative materials including transparent armor for windshields and visors. Armor research has been and continues to be a significant activity at the Army Research Laboratory.

Other subjects of significant interest are those related to ordnance materials, including propellants, projectiles, and even the systems used to shoot them. One area of concern lately has been to find replacements for lead bullets and depleted uranium (DU) kinetic energy penetrators. Environmental concerns are the primary reasons for finding alternative materials for these applications, and several of the articles here discuss the programs addressing the problem.

One approach to reduce the weight and complexity of systems is to develop multifunctional materials that perform two or more primary functions. Army researchers have many programs underway that are leading to technologies that exploit this concept and several of them are mentioned here. A multitude of other technology development efforts are also being examined to develop the new generation of lighter, higher performance materials needed to improve warfighting effectiveness.

The twenty separate articles contained in this issue of the *AMPTIAC Quarterly* will provide you with a glimpse at some of the technologies that will enable the Army to transform into a more mobile, survivable, and lethal force while simultaneously becoming a better steward of the environment. There are numerous technical challenges yet to be overcome, but as the reader will notice the Army Research Laboratory's Weapons and Materials Research Directorate (ARL/WMRD) is actively pursing those technologies necessary forthe Army to transform the face of the new battlefield.

Nanomaterials technology offers tremendous potential for reducing weight and increasing performance of the materials used in Army applications. While nanotechnology, and specifically nanomaterials research, are still nascent in their development, there are trends emerging and applications becoming realised.

The nanomaterials technology discussed here and in the following article "Small Features Enable Big Results: Nanotechnology Research Uncovers New Multifunctional Materials for Future Army Systems" is focused on synthesizing and processing materials that exhibit unique and often tailored properties at the nanometre scale (1-100 nm). Characterizing, controlling and utilising these nanoscale properties in practical ways are some of the challenges that scientists and engineers face in developing nanotechnology to solve relevant problems for the Army. It is a challenge for scientists, for example, to modify materials at the nanoscale with great uniformity without using complex, time-consuming methods such as nanolithography.

Ongoing nanomaterials research within the Army includes the synthesis and assembly of new nanomaterial building blocks, as well as the development of nanoscale characterization and metrology tools that can probe the structure and response of nanoengineered materials. Potential Army uses for nanomaterials can broadly be classified into structural and functional applications. Inorganic and organic nanomaterials are being explored in lightweight structural applications, and potentially for personnel and vehicle ballistic protection concepts. The ability to control the microstructure and response of materials at the nanoscale can lead to new mechanisms for load bearing and energy dissipation. As seen in the article on composite armor in this issue of the *AMPTIAC Quarterly* one such effort involving the application of nanotextured fibre coatings for composites applications is described. Likewise, transparent polymer nanocomposites are being explored for use in transparent armor and scratch-resistant transparent coatings.

Functional nanomaterials can improve performance of existing applications or enable new uses or capabilities. These two articles describe several research activities involved in developing nanofunctional materials and understanding and utilising their unique properties. They further explain how these properties may be exploited to address the needs for new Army materials.

Nanofunctional Polymer Materials

The fundamental units of organic materials are ideally suited for the development of nanofunctional materials. Synthetic chemists have developed numerous tools to create new nanoscale materials from the "bottom-up." Recently these tools have been applied to create materials with unique and desired characteristics and properties at the nanoscale. Nanofunctional organic materials are of particular interest to the Army, with specific examples described later in this article.

Military Uses of Nanotechnology and Converging Technologies

It is difficult to predict what the global technology landscape will look like in 20, 10, or even 5 years into the future. But it is not difficult to predict that as advances are made, so too will the opportunities for misuse. This chapter summarizes information on emerging technologies that are expected to have significant economic, societal, and dual-use risk impact in the near future. As highlighted during the workshop, prominent among these are advances in knowledge and delivery technology that have increased the dual-use potential and risk of non-lethal bioregulators; and the convergence of nanotechnology and biotechnology in the form of DNA nanotechnology.

Nanotechnology (NT) is linked to many kinds of hopes– and fears – due to its potential for radical change in how we live, work, communicate, even who we are, in particular through the convergence with other technologies and sciences, such as biology, computer science, cognitive science, social science and more ("converging technologies" (CT), Roco/Bainbridge 2003, Roco/ Montemagno 2004, High Level 2004). In order to understand benefits and risks and to help in steering NT development and use, technology-assessment institutions have begun to study NT; in Germany, e.g., the Office for Technology Assessment of the Federal Parliament (TAB) has published a major study in 2003 (Paschen et al. 2003).

If NT will lead to an industrial revolution in general, this will affect all areas of technology use. A specific sector is formed by the armed forces; in the industrialised countries they have used new/high technology since many decades to achieve victory in armed conflict. Even though it is well known that the armed forces prepare for violence and destruction, including mass destruction (e.g. in the official and unofficial nuclear-weapon states), military use of new technologies is not often included into technology assessment.

One reason may be that war is seen as extreme exception so that the impact of military technologies on society are minor. This is problematic, however: First, military technologies could have direct or indirect effects on civilian life even in peace time. Second, if new military technologies would make war more likely, the affected societies would suffer more. Since military and civilian technologies increasingly overlap and since there are various direct and indirect interactions between both sectors, technology assessment of NT should systematically include military uses (and misuses). Control of technology development and use should comprise both, the civilian and military sectors, to avoid undermining of limits.

In a first systematic perusal of potential military uses of NT, I have found several applications that could produce serious dangers – for international security, for security in societies or for human nature (Altmann 2004, i.p.). In the later sections, I describe these applications and present recommendations for their preventive limitation. In order to set the stage, I start with general remarks on the military and its use of high technology and cast a look at military research and development of NT in the USA.

Military Use of High Technology

While superior technology has been a decisive factor in wars in all of history, it was only in the last century that the armed forces incorporated science into their organisation in a systematic way. In particular after World War 2, rich/large/leading nations built up big centres for military research and development (R&D) and funded innovation in military industry on a large scale. Much of this was connected to nuclear weapons and their carriers, but increasingly was carried over to all other areas of warfare. Armed conflict is about breaking the will of an opponent by violent force. The law and rules of civilian society – that exclude pursuing one's will by force – no longer apply. Only the law of warfare poses some limits on how force is used; in peace time, agreed arms limitations can reduce the likelihood of war and restrict the number and quality of its tools. Success in armed conflict depends on many factors, among them technology, strategy, morale and surprise. Since highest national values may be at stake, there is an immanent tendency to circumvent the rules in order to prevail. In peace time, there is a motive to covertly compile more armaments than what arms-control agreements stipulate. This emphasises the importance of verification of compliance. In actual conduct of war, on the other hand, compliance with the international law of warfare depends on the spirit of a force; public scrutiny, experience with rule-keeping by the opponent, but also the fear of reprisal play a role. When armed conflict is raging, there is no reliance on the law of warfare; the necessity of prosecuting war criminals by international courts is only too obvious.

Since technology increasingly provides advantages in battle, all potential opponents have strong motives to stay in the lead, or at least to keep the lag small. New fields of technology make possible new ways of destruction. Military exploitation of new technology means putting the new possibilities into practice: researching the potential, developing and testing devices. Sometimes international law or arms-control treaties impose limits– e.g. the Biological and Toxin Weapons Convention of 1972 prohibits all non-peaceful uses of biological agents and toxins, including development and testing. Offencive biological weapons are prohibited thus, whereas defencive protection (such as gas masks, vaccination) is allowed. However, in principle an opponent might break the rules covertly and achieve victory in this way. Thus, there is a strong motive to understand what such an opponent could bring to bear in order to be prepared to counter such development.

As a consequence, even a defencive-minded state can feel a need for some R&D of the forbidden offencive weapons. In such a way, scientific-technical advance has a tendency to undermine agreed limitation treaties. The only ways to contain this danger between potential opponents are 1) reliable verification of compliance with existing agreements, and 2) early conclusion of treaties for new technologies, that is preventive arms control.

A central means of securing effectiveness in combat is secrecy. Knowledge of technical details, strategies, plans, even the motivation of an opponent's armed forces makes attacking them easier, thus there is a strong motive to keep such information secret on the own side– and to gain it by espionage on the others. There can be many additional motives for secrecy, and states differ widely how they handle it. The more secretive an armed force is, the higher is the possibility of a covert breakout: massing of weapons or personnel, introduction of qualitatively new weapons so that one side would suddenly dominate militarily. This problem becomes more urgent if new technology promises revolutionary change, as is the case with NT. Again it points at the necessity of supporting agreed limits by reliable verification, but it also emphasises the need to balance the required transparency with the (perceived) need for secrecy.

When potential military opponents prepare for the possibility of war by using new technology, small differences in technology can lead to high differences in the outcome. It is relative differences that count mostly. Shooting somewhat faster, at somewhat longer range, looking somewhat farther etc. can all mean that one's systems survive whereas the other's are destroyed. This explains the race for the "leading edge" and the high price that the military are prepared to pay for new technology. It makes agreement on limitations more difficult. However, there are additional factors; e.g. a more defencive orientation can give more tolerance.

Mainly because of their very task – prevailing in armed conflict by selective or massive destruction – the armed forces want to prepare many of the bad and ugly uses of new technologies that civilian society will try hard to preclude. Criminal development of such prohibited uses – new poisons, eavesdropping equipment etc. – within society is severely restricted. Access to laboratories, expertise, testing grounds etc. is very difficult if one has to work clandestinely, under permanent threat of prosecution. On the other hand, in military R&D such activities are carried out in an organised manner under the protection and with the resources of the states. This sets a conceptually different framework for technology assessment and subsequent regulation.

Within societies, there is a higher authority with the monopoly of legitimate force that protects sub-state actors from attacks by others and prosecutes perpetrators. The international system, on the other hand, can still be characterised by the security dilemma. There is no overarching authority with a monopoly of force – states try to make themselves secure by their individual armed forces. Usually, strengthening the latter increases the mutual threats, leading to motives for further strengthening. If unchecked, the net outcome is decreased security of all.

There are several ways out of this security dilemma– they are mutually reinforcing. One is voluntary agreement limiting the mutual armed forces (arms control). Since the armed forces keep their task of prevailing in war, should it occur, there is an obvious friction between the motives for limitation and the motives for combat effectiveness. Nevertheless, an enlightened view of national security should lead to the preference of mutual limitations. A second alternative is restructuring of the armed forces so that combat efficiency is high in static defence of the own territory, but is low in attacking others. A third way out of the security dilemma is political détente, economic cooperation etc. so that one could not gain from an attack and it would become politically/culturally unthinkable. Finally, states could develop the United Nations to the international authority with monopoly of legitimate force similarly to the one that governments represent within states.

Military R&D of NT in the USA

After the Cold War, the USA has remained as the undisputed leader in military high technology. The USA spends about two thirds of the global expenditure for military R&D, $ 52 billion of about $ 80 billion in 2002, and marked increases are planned (BICC 2003: 48, 2004: 54). In NT proper, one quarter to one third of the Federal R&D money in the National Nanotechnology Initiative (NNI) goes to the Department of Defence (2003: $ 243 million of $ 770 million) (Altmann i.p.: Section 3.1). (There is not yet a special programme for CT.) This money is used for a great variety of activities, mostly at the stages of basic research, applied research and advanced technology development. Work is being done at universities, the laboratories of the armed services and the national weapons laboratories, on topics such as carbon nanotubes, magnetic nanoparticles, organic light-emitting diodes for displays, bio-computing, bio-molecular motors. Closer to military needs are projects in sensors for chemical- or biological-warfare agents or nanostructured explosives and armour. Development of systems that could actually be deployed is still many years off. In an effort for more targeted research and earlier introduction into service, the Army is funding the Institute for Soldier Nanotechnologies that was founded 2002 at the Massachusetts Institute of Technology, with $ 50 million over five years (plus additional $ 30 million from industry).

Here, up to 150 people are working in seven multidisciplinary teams. One overarching goal is a battle suit that protects against bullets, chemical and biological agents, provides cooling or heating, monitors body functions, compresses wounds and applies medicines.

In conferences of the NNI, various goals were given for "National Defence" or "National Security" uses of NT and CT (Roco/Bainbridge 2001: ch. 2, Roco/Bainbridge 2003: section E). Among them are:

- miniature sensors, high-speed processing and communication,
- sophisticated virtual reality systems for training,
- uninhabited combat vehicles,
- higher performance in military platforms,
- improved chemical/biological/nuclear sensing and casualty care,
- improved systems for nuclear non-proliferation monitoring and management,
- devices for control of nuclear systems,

- enhancement of human performance,
- a brain-machine interface.

Other countries are spending much less for military R&D of NT. A cautious estimate is that the US expenditure is four to ten times the amount of all other countries combined. However, this could change when other states that are active in military high technology would follow the US example and make military NT a matter of high priority.

By leaving to the USA the task of trying many avenues and sorting out the more promising ones, such countries could not acquire a leading position, but might nevertheless get relatively more military effect for the same amount of R&D spending. There is little doubt that countries which are capable in NT in general could also develop and introduce military NT applications, following the US lead with several years delay.

Potential Military NT Applications

NT could be used in all areas of fighting and preparing war. Small but powerful computers will be built into weapons, uniforms, equipment, vehicles etc. Larger computing systems will be applied in battle management and logistics, increasingly taking over tasks that used to be reserved for humans. Small, sophisticated sensors could become cheap enough to be scattered in high numbers. Missiles will become smaller, guidance systems could be provided even for small munitions. NT-based fibre composites could allow metal-free small arms and ammunition. Vehicles will become lighter and more agile, engines more efficient.

Autonomous vehicles for all media will become possible, for reconnaissance and communication, but also for combat. For outer space, satellites and their launch rockets could become much smaller and cheaper; they could form anti-satellite weapons working by direct hit or manipulation after docking. Some autonomous systems may be very small (centimetres, later millimetres and even below); if fully artificial systems will turn out more difficult to construct, modifying/implanting insects or small mammals may provide an earlier solution. Providing neurone contact, bio-compatible materials, small computers and power supplies, NT could be used for implants in soldiers' bodies– for monitoring, communication, expanded senses or release of drugs.

Genetic engineering and biotechnology already *per se* provide options for new chemical and biological warfare agents that are resistant against antibiotics or suppress the immune system (e.g. Nixdorff et al. 2003). NT will likely accelerate such possibilities and create new ones, generically by new analysis and manipulation tools. Directly utility will come with the advances that will be made in individualised and targeted pharmacology, such as nanoparticles for ferrying agents across the blood-brain barrier, organspecific binding or *in-*

vivo recognition of DNA patterns. The latter may lead to selective action, removing much of the traditional military hesitation towards biological agents because the own forces and population could be infected, too. On the other hand, NT-based sensors and filtering/neutralising materials could be used for better protection against chemical or biological agents.

Preventive Arms Control: Concept and Criteria

In order to provide a framework for assessing militarily relevant technologies, the concept of preventive arms control has been developed. The purpose of preventive arms control is to limit technologies before they are deployed with the armed forces, often starting at the development or testing phases. The concept proceeds from the viewpoint of international security, looking at the international system – and is thus different from an outlook that tries to provide national security through military strength. However, if national security is seen in the wider context, this may well lead to acceptance of preventive limits of the own technological capabilities if those of potential opponents will be reliably limited likewise. Earlier examples include the Anti-Ballistic Missile Treaty (1972-2002) or the Protocol banning Laser Blinding Weapons (1995).

In today's situation, the concern of preventing access by terrorists to certain technologies should provide additional motives for preventive arms control. Of course, terrorists cannot be partners of limitation agreements, but their own capabilities of making use of new technologies are quite limited; much more likely is access to technology that would have been developed and produced by states.

Beside agreed limitation (arms control proper) – earlier in many cases bilateral, then increasingly multilateral –, there are also unilateral steps that can be taken. If the situation is very asymmetric, a technology leader can renounce a certain principal possibility without danger. As long as comprehensive agreements will not be in effect, export controls by the most advanced nations – a form of multilateral unilateralism – will be useful to limit the spread of new military technology to countries that may not be inclined to enter agreements. However, export controls are discriminatory and create motives for circumvention – universal agreed limitation is clearly preferable, even if a few outsiders may remain.

Preventive arms control consists of four steps:

(1) prospective *scientific-technical analysis* of the technology in question;

(2) prospective analysis of the *military-operational aspects*;

(3) assessment of both under the *criteria of preventive arms control*. If the latter leads to the result that action is recommended,

(4) possible *limits* and *verification* methods need to be devised. The criteria can be sorted in three groups:

I. Arms control, disarmament and international law:
- Prevent dangers to existing or intended arms-control and disarmament treaties,
- Observe existing norms of humanitarian law,
- No utility for weapons of mass destruction.

II. Stability:
- Prevent destabilisation of the military situation,
- Prevent technological arms race,
- Prevent horizontal or vertical proliferation/diffusion of military-relevant technologies, substances or knowledge.

III. Protect humans, environment, and society:
- Prevent dangers to humans,
- Prevent dangers to environment and sustainable development,
- Prevent dangers to the development of societal and political systems,
- Prevent dangers to the societal infrastructure.

The readiness to enter preventive arms control and the extent of limitations will be influenced by the military posture of the respective state. The tasks given to the armed forces may differ widely, the spectrum may range from large-scale armed conflict everywhere on the globe via defence of the own territory, crisis intervention and peace enforcement to defence against terrorist attacks. With its priority on international security and agreed limitations, preventive arms control is inclined towards reducing military threats and offencive potential. It makes sense on all levels of the military-posture spectrum. Without it, the more offencive postures could become more dangerous over time.

With preventive limits in effect, there is the principal possibility of clandestine breakout. This has to be assessed in advance and precautions have to be taken. In the technology areas regulated specifically, this can be done by appropriate rules and methods of verification of compliance.

Since science constantly moves into the unknown, the advances need to be monitored continuously, e.g. by an international body comprising scientists, military and disarmament professionals. Wherever problematic military applications appear on the horizon, military-technology assessment should start and adequate specific limits with appropriate verification should be developed, if needed.

Such an approach could also solve the problem of dual use; civilian R&D needs to be included from the beginning into monitoring, assessment and, if necessary, limitation.

Since states have to enter arms-limitation treaties voluntarily, they will only do so if they are convinced that the agreements are in their best interest. In pondering the arguments for and against, not only the military should be heard; scientific analysis and an enlightened public should play an important role.

Assessment of NT under Criteria of Preventive Arms Control

In my assessment of the potential military NT applications under the criteria of preventive arms control (Altmann i.p.: Ch. 6), I have found several that do not pose strong dangers or that are so close to civilian uses that limitation would be impractical. This concerns the more generic applications such as computers or materials. A few military uses could help to protect against terrorist attacks (sensors, filters, decontamination agents) or would improve the verification of existing treaties (certain sensors). These are rather judged positive.

On the other hand, there are several potential military NT uses that raise flags in the criteria groups:

Arms Control, International Humanitarian Law: New types of chemical or biological weapons would violate the Chemical Weapons Convention (CWC) or the Biological and Toxin Weapons Convention (BTWC). R&D in these directions could undermine these conventions over time. Particular problematic is the fact that the BTWC – different from the CWC – that does not yet include a compliance and verification protocol.

NT-enabled lighter combat vehicles could evade the definitions of the Treaty on Conventional Armed Forces in Europe (where, e.g., a "main battle tank" is defined to have a mass above 16.5 metric tons).

Autonomous combat vehicles – seeking their targets by themselves and attacking them – would violate the international law of warfare because they would for quite some time not be able to differentiate between combatants and non-combatants or to recognise combatants *hors de combat*.

Stability: Military stability between potential opponents could be decreased by NT-enabled small, highly precise weapons that enhance offence. Small missiles (size maybe 0.3 m) could be cheap and used – against aircraft, communication nodes etc. – in high numbers. Particularly worrisome would be small ballistic or cruise missiles (say, 1 m length) that due to their extreme precision (say, 0.1 m accuracy) were capable of attacking nuclear-missile silos and strategic command centres with a non-nuclear payload.

Nervousness would rise with micro-robots that could covertly invade military systems and strike from within at any time. Autonomous combat systems at close mutual range (at sea, in outer space) would need to react fast, creating the possibility of uncontrolled escalation. Arms races are likely in all areas of military NT use.

Technologies for light vehicles, small missiles, precision guidance, biological agents etc. will spread. If NT-based military systems will become small and cheap, proliferation will be difficult to control.

Humans, Environment, Society: Dangers to humans and society in peace time would arise if new chemical or biological warfare agents would get into the hands of criminals and terrorists. The definition could of course be updated easily by the Treaty Parties as new light-weight combat vehicles would arrive. The main problem would be to ensure that this actually happens and is not blocked by states that hope for military advantage, maybe outside of Europe. The "smart" weapons of today are a step into this direction, but in their case the target has been defined and the firing decision made by a human operator. The target is still found in a relatively simple manner – e.g. by fixed co-ordinates (cruise missile), with human-operated designation (laser-guided bomb), by special vehicle signature after arrival in a human-specified area (anti-tank submunition). Small arms and munitions made from nano-fibre composite without metal would evade existing x-raying and metal-detection equipment, making criminal and terrorist attacks easier. NT-based materials, propellant and guidance may allow missiles of less than 0.5 m size that could take civilian airliners down.

Privacy could be invaded if small sensors or robots, originally made for the military, would make their way into society. Should the armed forces rush ahead with manipulation of soldiers' bodies, civilian society would be faced with *faits accomplis* that would pre-empt the needed fundamental debate how to deal with the potential of modifying the human condition.

Such systems could be much more easily hidden and transported than present so-called man-portable air-defence systems (MANPADS), such as the US Stinger with 1.5 m length and 16 kg mass. Having provided Stinger systems to the Mujahideen fighting against the Soviet Army in Afghanistan, the USA has later carried out a buy-back programme with limited success.

Recommendations for Preventive-Arms-Control Action

Since there are strong reasons for concern, certain military applications of NT should be prevented. At the same time, positive civilian applications should not be hampered unduly. However, civilian society will have many reasons to block or control problematic NT applications anyway. Even though this is superficially independent of preventive arms control, there are clear interactions between civilian and military R&D, and the dual-use potential will grow as NT will advance towards manipulation of matter on more fundamental levels. Thus, civilian limits should be co-ordinated with military ones. This calls for an internationally harmonised approach.

Potential limits were considered systematically, together with the effects on benefits of civilian uses and the requirements on verification of compliance. As a result, no rules tied to NT as such are recommended – NT is just too broad. Rather, specific military missions or classes of equipment should be limited, irrespective of the technology used inside. This follows a tradition in arms control: it is systems and missions that can act destabilising, not technology per se. A second important argument is about verification. Types of equipment designed for specific missions can usually be recognised from the outside; technology-specific limits, on the other hand, would require deep intrusion into the military systems which would be unacceptable.

The regulation for applications of NT should be embedded into the already existing framework of arms-control and disarmament agreements and international humanitarian law. Specifically, I recommend the following measures:

- Small, self-contained sensor systems below a few cm size should be banned. Smaller sensors are not limited if they form components of larger systems. For important non-military uses, few exceptions might be agreed on, e.g. "bugs" for legal eavesdropping by law-enforcement agencies.

An example would be a fictitious rule not to use carbon nanotubes (as high-strength material, as elements of future high-density computers) in military systems.

However, since that NT will make possible small and very small systems, the intrusiveness will have to increase if such systems will arrive. This is another reason for the macroscopic size limits in my recommendations. Even so, microscopes of various types may have to be accepted as on-site-inspection tools. Verification of the Biological and Toxin Weapons Convention will of course need laboratory visits, sample taking, and other microbiological techniques with or without NT.

- Metal-free small arms and ammunition as well as small missiles (below about 0.5 m size) should be banned.
- There should be a ten-years moratorium on body implants and other body manipulation that are not directly medically motivated.
- Pilot-less combat aircraft and other mobile armed combat systems without crew should be banned, with numerical limits on autonomous systems that are unarmed.

- Small robots (below 0.2-0.5 m size) should be banned, with narrowly limited exceptions (e.g. for exploration of collapsed buildings).
- A comprehensive ban on space weapons should be concluded; small satellites (below, say, 1 m size) for other purposes (e.g. earth monitoring or satellite maintenance after docking) – that could nevertheless be used to hit or manipulate other satellites – should be limited in number.
- The Chemical Weapons Convention should be upheld.

The Biological and Toxin Weapons Convention should be strengthened by a compliance and verification protocol. Where necessary, clarifying interpretations should be added. These limits should hold for the civilian and the military sector and start at the stages of development and testing, since it is here where the purpose becomes visible and verification feasible. Deployment and use should of course be prohibited, too.

In order to not create too much resistance to such a new rule, existing guided one-way weapons such as cruise missiles should not be affected. The proposal is to use the word "re-usable" in the definition. Since "autonomy" cannot be determined from the outside and could be introduced by a software change, this category should not be used in the legal definition. Differentiation from mines is achieved by the notion of mobility.

If it will turn out that re-usable, armed, crew-less systems cannot be prevented (the US drone Predator, retrofitted with a Hellfire missile, has already been used under remote control to destroy a car and kill six putative terrorists in Yemen in 2002), then at least it should be agreed that aiming and weapon release should always be done by a human in the decision loop and that qualitatively new types of nuclear-weapon carriers (such as robots) are banned. The definition should include artificial systems as well as hybrid artificial-biological ones, e.g. rats or insects with implanted control electrodes. Including updating of the schedules of banned and controlled chemicals, as required by scientific-technical advance.

In the CWC this would concern potential damaging/incapacitating processes (e.g. electrical, thermal ones) for which some might argue that they do not constitute "chemical action" as required in Art. II. In the BTWC this is about potential (semi-)artificial agents where arguments might be made that they are not "microbiological or other biological agents" as required in Art. I.

In a simplified linear model of a military technology, the life-cycle stages are: research, development, testing, production, acquisition/deployment, use, taking out of service. Because during the research stage the later application is usually unclear, preventive limits often exclude research.

With the choices of several centimetres to several times ten centimetres as the lower size limits, verification of compliance can rely mostly on on-site inspections with observation from outside. For chemical and biological agents, however, taking of samples and special analysis equipment will be needed.

The recommendations are geared towards mutual limitations that are agreed internationally by states and will thus affect all partners (more or less) equally. By foreclosing many avenues of production and transfer, they would also have an asymmetric effect on non-participating states and non-state actors – the stronger, the more comprehensive the participation would be. Before negotiations might start and while they might drag along, co-ordinated restraint among the states most active in NT – by tacit or explicit agreement – would be useful. This concerns own development of problematic military systems as well as exports of systems and technologies.

These limits should be co-ordinated with the regulation governing NT use within states as well as with the code of conduct on responsible R&D and use of NT that is beginning to evolve (International Dialogue 2004). The discussions about such a code of conduct should explicitly include the military field.

In order to make the recommendations more concrete and practicable, further work on definitions and delineation as well as on verification details is needed. The debate on limits of military NT can be much alleviated by transparency in NT R&D. In particular, those states that are traditionally not open about their military R&D should re-think their posture. Restraint in military NT R&D can also help, in particular concerning potential offencive or otherwise threatening uses. However, verification of development and testing limitations will usually have to look at research establishments, too.

An Implementation Strategy for Nanotechnology

Because nanotechnology represents a new and dramatically different approach to manufacturing and utilisation, it would behave the Army to enlist public support from the earliest possible time. To the maximum extent possible, an effort should be made to involve stakeholders to avoid issues similar to the lingering and unsubstantiated concerns surrounding genetically modified foods. Thus, the Army should encourage citizen participation in the development of nanotechnology for military applications. Army acquisition strategy, through the Acquisition, Logistics and Technology infrastructure can be used to shape the development of nanotechnology in a manner both protective of the worker and soldier as well as protective of the environment. The Army Environmental Policy Institute (AEPI) can provide guidance and oversight of the safety and environmental aspects of the acquisition requirements and process.

The Army, and specifically AEPI, can influence the progress of nanotechnology through a carefully crafted acquisition strategy. For example, the Army can demand high standards for testing new materials. Developmental shortcuts may be tempting, but the goal must be the safe advancement of the technology without the lingering environmental disasters that were experienced with the manufacture of nuclear weapons. Similarly, the Army must insist that contractors must comply with the recognised best practices in the nanotechnology industry.

What Lies Ahead: Predictions

There is a great deal that remains to be learned about nanotechnology. Because the materials have new and rather different properties, a whole new set of manufacturing challenges will emerge. It will require time, effort and funding to resolve these challenges. Particular areas of challenge include the ability to control nanodevices from the macroworld as well as the altered nature of physics when the surface to volume ratio of material is changed to the nanoscale.

Our competitors, both nations and NGOs will continue to develop nanotechnology; regardless of the ethical implications of nanotechnology, the technology will progress; for example, hundreds of tons of nanomaterials were synthesized in the United States in 2003 and within a decade, the market in these materials is predicted to exceed $1 trillion.

Because the basic elements to initiate a nanotechnology research program are relatively simple and because the world's economy is becoming increasingly globalized, nanotechnology will develop simultaneously in a number of different locations and in a number of different directions. This will make it difficult to monitor developments in the technology and will make it especially difficult to detect efforts to weaponize the technology. Nevertheless, it is vital that the potential for harm from nanotechnology be identified and tracked. Additionally, it is essential that links between research and development nodes with a nanotechnology focus and manufacturing facilities be carefully monitored.

Nanotechnology will cause major changes in our way of life. In the same way that night-time progresses through dawn to daylight, the changes wrought by nanotechnology will be gradual. It is inevitable that nanotechnology will be weaponized and employed against the United States. The form that the employment will take remains to be defined, but nanotechnology will offer radical new changes in the way in which we wage war at both the strategic and tactical levels. The changes could include effects identical to recognised chemical weapons or small nuclear weapons or attacks on food supplies or even devices to attack particular age groups or populations. It will be necessary for the United States to develop countermeasures to nano weapons.

Recommendations

The Army leadership should develop a specific focus on nanotechnology. Because the technology has uniquely transdisciplinary characteristics, the observers must be prepared to incorporate principle of chemistry, physics, biology, computer science and ethics as the technology matures. In particular, the effort should take pains to distinguish the real applications of the technology from the imaginary. For now, anyone who delves into the potential of nanotechnology must assume that they don't know what they don't know.

a. The Army should develop a specific focus on the implications of nanotechnology for land warfare. The development of such a focus will likely require the input from a variety of disciplines including electrical engineering, biomedical engineering, and environmental sciences. The potential use of nanomaterials by opposing forces must be a clear interest.

b. There should be continued research on the ESOH effects on nanomaterials on the soldier and his/her environment. Particular attention should be paid to the biological effects of nanomaterials, both beneficial and harmful. The continued research in nanotechnology must be seen favourably by the public. There is a danger that the pubic may misperceive and exaggerate the risk of the technology such that nanotechnology is viewed as a fatally flawed technology such as genetically modified foods or nuclear power.

c. There should be the identification of the predictable consequences and an effort to uncover the unintended consequences of nanotechnology.

  - Special attention must be given to aspects of the technology which may impact public health and the environment.
  - Within the workplace, the Army must promote the recognised best practices of the industry in terms of risk assessment and environmental safety and security.
  - Careful surveillance of the nanotechnology workplace to identify known and new forms of occupational diseases; in particular efforts should be made to gather data on worker and soldier exposure, toxicology and dose response.
  - The Army should cooperate with other stakeholders in the development of standardized risk assessment tools and appropriate regulations.
  - To the maximum extent possible, the Army should strive to minimize the production and release of nanoparticles into the environment.

d. Determine if nanomachines are chemical weapons under the provisions of the Chemical Weapons Convention.

e. Include the applications of nanotechnology in the war gaming scenarios used to train Army leaders.

Nanotechnology and National Defence

The current and future world security environments offer many challenges to the United States; some of the most formidable challenges will be found at the intersection of technology and radicalism; the magnitude of these challenges is only amplified by the increased mobility of modern society.

Our recent successes and current technological domination must not be allowed to lull us into a sense of false security. The global proliferation of nanotechnology and nanotechnology based weapon systems will change the face of conflict and it is critical that our nation maintain its technological superiority. The effect of nanotechnology on weapons systems remains to be defined. On the one hand new and more destructive weapons are possible using nanodevices, but on the other hand, nanotechnology may make the production of conventional weapons so cheap, perhaps as little as a dollar per pound, that the threat will be derived from the massive production of traditional weapons.

Nanotechnology represents both a challenge and an opportunity for the national defence. The National Military Strategy of the United States of America (2004) accurately notes that technology is one of the key determinants of the security environment particularly the diffusion of and ease of access to technology by potential adversaries of the United States and its allies. In response to the threats present in the national security environment, the document cites agility and integration as two of the principles guiding the development of the Joint Force as it seeks to attain its military objectives: protection of the United States against external attacks and aggression, prevention of conflict and surprise attack and to prevail against adversaries. To sustain and increase the existing military advantage held by the United States will require a transformation of traditional and parochial mindsets; this transformation will require a combination of technological, intellectual and cultural changes across the joint community.

Nanotechnology as a Challenge

Nanotechnology may find fertile ground for growth at the intersection of radicalism and technology. Information about the developmental aspects of the technology is freely available in scientific journals, seminars, etc. and is available to various state and non-state adversaries. The methods for processing nanomaterials do not require the substantial infrastructure required to manufacture conventional weapons, thus nanotechnology is well suited to be an asymmetrical weapon. Indeed, nanotechnology may be more than just a breakthrough technology in the way that development of transistors allowed breakthroughs in the electronics industry; nanotechnology may be better described as leap ahead technology because the nations or organisations that dominate the technology in its developmental stages may come to dominate other entities with a lesser grasp of the technology. Because of its significance nanotechnology must not be allowed to develop and to proliferate beyond our sight.

Applications of nanotechnology may threaten the United States with both disruptive and catastrophic challenges. The nature of the technology makes it virtually undetectable and it could be used, for example, to sabotage petroleum or food supplies. It readily lends itself to applications much like a silent but potent WMD against civilian population clusters or economic centres and it is not detectable through the usual means of detection for chemical or biological agents. As a new weapon against civilians, nanotechnology could have devastating and dramatic negative psychological effect on the American population. Nanotechnology as a tactical threat

Consider this scenario: A column of soldiers moves through the close confines of a city. Because of the potential for hostilities, the soldiers are maintaining a MOPP level 2 posture and chemical detectors are deployed in the column. Suddenly from the surrounding rooftops, there are gunshots and a number of canisters are hurled off the roof tops. Within moments, portions of the column are enveloped in hazy cloud and within a minute or so the soldiers closest to the canisters are twitching and salivating uncontrollably and even those soldiers who were able to don their protective masks and gloves are showing the same symptoms. Soldiers from the rear of the column move forward having easily cleared the roof tops with automatic weapons fire in an effort to aid their comrades.

Although the chemical agent detectors show no evidence of conventional chemical agents, they administer nerve agent antidotes in accordance with their training, but the victims worsen and quickly die. Within a few minutes, even the fully garbed soldiers find themselves salivating beyond control and trembling. Soon, they too are dead; the chemical agent detectors remain silent. What happened here is but one possible result of nanotechnology harnessed to do the will of terrorists. Traditional chemical agents are largely prohibited by treaty or agreement and the precursors of traditional agents can be tracked. As nanotechnology advances, it will be possible to design materials that act like chemical agents, in this case a cholinesterase blocking agent, but are not classed as chemical agents under any existing protocol, do not trigger existing chemical agent detectors and in any case do not respond to known nerve agent antidotes and, because of their small size, can penetrate protective fabrics and even mask filters.

Centres for the development of nanotechnology are scattered around the world and are limited by the availability of local talent and access to information; consequently there is little to impede these centres for development. Wherever there is a confluence of expertise in biology, physics, and chemistry and information technology, there is the potential to advance nanotechnology. Thus academic centres, research facilities and commercial organisations, as well as interested individuals, can develop nanotechnology threats. Dual use technologies should be a special focus of concern, particularly in the areas of pharmaceuticals, information technology and high resolution imagery.

As an example of the development of nanotechnology in countries of interest to the United States, consider Iran. The *Teheran Times* reported recently that a centre for nanotechnology was established in Isfahan province.

This is notable for a number of reasons. The Minister of Science, Research and Technology, Ja'far Towfiqi, claims that Iran is one of the world's leading nations in science and has budgeted, "Two hundred billion rials (approximately 25 million USD) for the development of nanotechnology, biotechnology and support of research...."

Even Iranian President Khatami included references to nanotechnology in his report to Ayatollah Ali Khamenei during his annual Government Week report of August 2004; the president claimed he had established a nanotechnology council under his own supervision 8-10 months before. The importance of a nanotechnology centre in Isfahan is further highlighted by the fact that the city is the home of the Isfahan University of Technology and is suspected to be the focal point of much of Iran's defence and research industries. The city is also the site of the Iranian nuclear program.

China represents another area in which advances in nanotechnology are of interest to the United States; not much is known about the Chinese effort. The Chinese are known to have an interest in post nuclear weapons and have established Project 863 which purportedly focuses on nano weapons. This project is thought to involve applications of nanotechnology designed to destroy or disable nuclear submarines. In addition, China has sponsored conferences on nanotechnology such as the one held in Beijing in November, 2002 with support from the People's Liberation Army General Equipment Detachment and the National Defence Science and Engineering Committee. A number of the presentations included discussions of molecular self assembly which is a key component of advanced nanotechnology.

The existence of nanotechnology will require new methods of deterrence as well as new operational methods should deterrence fail. Intelligence systems must be able to identify hotbeds of nanotechnology research among our adversaries. At a minimum, it must be possible to detect advances in nanotechnology that may offer a threat, identify the origin and significance of the advances and track the spread of technological advances as they become known to researchers. Because it is an emerging technology, the goal must not be to simply catalogue current capabilities, but to create a time line projecting future developments.

Nanotechnology as an Opportunity

The Nanotechnology Opportunity ReportTM [NOR] is the first worldwide overview of this burgeoning field from a business perspective. Profiling over 700 companies, venture capital investors and research institutes from around the world, the report provides key information for investors and executives to enable them to understand and take advantage of the myriad opportunities provided by this tiny revolution.

Nanotechnology can help the United States maintain a decisive qualitative advantage in the application of will. As it develops, nanotechnology will find applications in a variety of force concept areas. Within the realm of force application, the technology will enhance the explosive force of conventional explosives. When combined with nuclear weapons, nanotechnology enables the design of much smaller devices which, combine with precision guided delivery systems, can find wider applicability on the battlefield; it may be conceivable that weapons developers may advance from back pack sized nuclear devices to cigarette carton sized devices.

Nanomaterials have particular value in protective applications. For example, current efforts are underway to design body armor which has the look and weight of ordinary uniform material but, when an electrical field is applied, becomes many times stiffer offering ballistic protection. Similar materials may have the ability to change colour thus providing new methods of camouflage; even an illusion of invisibility may be possible. There are numerous possible applications of nanotechnology in the field of military medicine. Body sensors are envisioned which will permit the vital signs of individual soldiers to be tracked from a distance.

The technology will offer new drug delivery systems as well. At some point the technology may be able to offer a generalized immune booster thus obviating the need for a series of immunizations prior to deployment. In a similar fashion, such an immune boosting agent may have the ability to resist biological and chemical agents designed by an adversary to circumvent conventional medical defences. There is also great potential for nanotechnology to aid in the design of devices to detect hidden explosives, for example car bombs or improvised explosive devices.

Application of nanotechnology to the military logistics process will yield major benefits. Advances in nanotechnology will help to further miniaturize radiofrequency labels for materials; because the memory on chips constructed with nanotechnology is nonvolatile, the energy consumed by the radiofrequency labels will be minimized. Such labels will permit accurate tracking and stocking of supply items and will aid in just in time delivery to the end user. Advances in nanotechnology will also decrease the maintenance burden. New coatings will offer both better environmental protection and chemical agent resistance. Nanomaterials will offer superior lubricants and extend the useful life of machinery. There is an excellent possibility that nanotechnology will offer lighter and more energy efficient means to generate power in a wide range of settings from installations to tactical settings.

As an aid to battlefield awareness, nanotechnology will offer an unprecedented, and as yet hardly imagined, array of sensors. These will be very small, inexpensive and internally powered devices which can be spread across a battlefield by a variety of means such as artillery shells or unmanned aerial vehicles to offer early warning of an adversary's size, activity, capabilities and intentions. Such device could also be used in an unconventional or counter terrorism modality to track items or persons of interest.

When applied to command and control functions, nanotechnology may have its most immediate applicability. Within the visible technology horizon, nanotechnology will offer much smaller and more powerful communications devices and memory chips. Memory chips made with nanomaterials have the advantage of much denser and thus greater storage capacity and they also offer non volatile memory, thus the memory persists without consumption of electricity even when the device is not powered. The technology will offer the combatant commander a better view of his troop dispositions because the technology will enable much larger display screens with better resolution than now available.

Nanotechnology and the Joint Force

Two of the major principles of the Joint Force are agility and integration; nanotechnology will have a major impact on both of these concepts. Agility is defined in the National Defence Strategy as "the ability to rapidly deploy, employ, sustain and redeploy capabilities in geographically separated and environmentally diverse areas". As it develops, nanotechnology has the potential to greatly decrease the load carried by the soldier and the logistical tail needed to support a soldier in theatre. Within the foreseeable future, nanotechnology will help to miniaturize communications devices, make armor lighter and energy sources smaller and more powerful. A decrease in the load carried by the soldier will translate in faster movements both into and within theatres as well as fewer sorties and platforms required to move the force.

Integration is a concept which "focuses on fusing and synchronizing military operations among the Services". Because nanotechnology is in an early stage of development, it will be possible for the Services, other governmental agencies and non governmental organisations (NGOs) to integrate nanotechnology in a seamless fashion across a wide variety of applications. For example, the development of individual communications devices may be standardized thus decreasing both per unit costs and the number of repair parts required. Because nanotechnology is in such an early stage of development, it still has the potential to be ""born joint," i.e., conceptualized and designed with joint architecture and acquisition strategies."

Indeed, because nanotechnology is at such an early stage of development and will have such wide applications, both in an out of the Armed Forces, the joint development of the technology for military applications may serve as a vehicle to advance the concepts of a joint force with expeditionary capabilities. Development of new command control and communications tools using a common derivation of nanotechnology will help to break down the cultural barriers between Services and offer a common ground for joint planning and operations.

Nanotechnology, as it develops and flourishes will change the way warfare is conducted in much the same way that gunpowder brought about the end of the era of edged weapons. Indeed, nanotechnology will offer national decision makers a wider range of responses. Our current military technology emphasizes destructive force. Advances in nanotechnology will provide a range of response options from total destruction to total control. For example, total destruction is well understood and may be obtained by conventional means but total control may be obtained by employing nanodevices which neutralize an opponent's energy sources thereby freezing his communications and mobility. A mastery of nanotechnology, both it threats and benefits, is a key to maintaining unchallenged military superiority; such mastery will require substantial investment. Unfortunately some experts believe that the United States is all ready under funding nanotechnology as compared to other nations and predict that our failure to provide adequate funding will result in a loss of U.S. defence and economic leadership over the next 20 years.

Another proposed role for nanotechnology in weapons development is in fourth generation nuclear weapons. These explosives employ inertial confinement fusion methodology to create weapons with yields ranging from a fraction of a ton to tens of tons. When delivered by a precision guided device, such weapons will have devastation effects combined with comparatively small size and almost no fallout.

Current Army Effort

The current Army effort in nanotechnology is focused at the Institute for Soldier Nanotechnologies (ISN) housed at the Massachusetts Institute of Technology. The ISN was established in 2002 by a five year 50 million dollar contract from the United States Army and its primary goal is to create a "21st century battle suit" intended to enhance soldier survivability. A key goal of the program is to use nanoscale devices to decrease the weight of the soldier's equipment. For example, one concept under investigation is "dynamic armor"; a fabric containing nanomaterials is made many times stiffer when an electrical current is applied. Such a fabric could be used to offer body armor on demand or as an instant splint for an injured extremity. The institute uses a team approach to examine a wide range of projects. One team investigates energy absorbing materials while another team looks at mechanically active materials and another team examines sensing and counteraction devices.

The notion of a 21st century battle suit is part of the Army's effort to transform itself into a campaign quality force capable of deploying with much greater agility; nanotechnology is seen as offering dramatic improvement in the mobility of the individual soldier and small units. In particular, the ability of nanomaterials to shrink the volume and weight of a soldier's equipment will aid in the rapidity of deployment.

What are the next steps? Nanotechnology will likely cause major changes in the social, economic and political dynamics of the United States. The magnitude of these changes will mandate a national strategy as well as an Army strategy to insure a smooth transition to a new technology.

One of the keys steps in a national strategy will be the development of the educational resources needed to support a nanotechnology industry to include basic educational skills and specific technical training programs. Unfortunately, the number of American graduates with advanced scientific degrees is dwindling while the numbers of PhD prepared foreign graduates from American universities is increasing. Attainment of an adequate number of graduates in the scientific disciplines may require commitment to scientific education analogous to the commitment seen in the United States after the launch of the Sputnik.

A flourishing nanotechnology industry will bring a great deal of wealth to an as yet to be determined number of people. As the industry develops in the United States, efforts must be made to insure that the benefits of the technology are shared by all citizens. Major social disruptions were seen as the industrial revolution advanced; such disruptions may occur as nanotechnology advances. To the extent that society is capable of learning from past experiences, we should strive to see a more equitable distribution of the wealth created by nanotechnology.

An Army strategy will develop as an outgrowth to the national strategy, but the Army strategy will be oriented toward the application of nanotechnology to military purposes. Predictably, since the Army draws from the civilian populace, Army strategies will spill over into the civilian concepts about nanotechnology and thereby help to shape the expectations of Americans regarding the technology. The Army's strategy must be keyed to both tangible milestones as the technology evolves and to the important goal of establishing nanotechnology as a sustainable process.

Reasonable and tangible milestones for the development of the technology could be based on the criteria of the National Institute of Nanotechnology (Canada). At first, simply the assembly of devices on a liquid surface would be achieved, and then followed by the application of a power source, perhaps derived from ATP or some other chemical source.

Subsequently, it would be reasonable to expect that the nanodevice be able to recognise changes in its surroundings. An advanced device should be able to perform specific elective functions as determined by a user.

These are not intended to be all inclusive milestones, but are examples to demonstrate how progress in nanotechnology could be monitored.

3

Societal and Defence Implications of Nanotechnology

Societal Implications

Just as the industrial revolution began with the mundane need to pump water out of coal mines, the initial applications of nanotechnology provide us with improved tennis balls and better cosmetics; like drier coal mines, these modest benefits are hardly the foundation stones of an historical shift. But just as the steam engine progressed to the internal combustion engine and then far beyond the needs of the coal industry to a society with unprecedented mobility, nanotechnology will have profound societal implications. At this point in time, any discussion of the societal implications of nanotechnology is purely speculative, however, if based on a set of assumptions described here, some likely outcomes may be predicted and speculative societal implications can be derived from the proposed outcomes.

Assumptions

There are a number of reasonable and realistic assumptions that can be made about the future success of nanotechnology. First, this scenario is looking up to 25 years and beyond in the future. It seems likely that this is a long enough interval to allow nanotechnology to advance to the point that nanoscale machines will be employed in a systematic fashion to construct atomically precise products at very low cost and with a minimum of waste material and with much less energy and extraneous heat. The machines, in aggregate, will be powerful and able to produce massive quantities of products at low cost, probably in the range of a dollar per pound, exclusive of development, legal, insurance, environmental, safety and marketing costs.

In essence, industrial society will reach a point where it will be possible to arrange atoms in an efficient and economic fashion limited only by the laws of nature. It is likely that there will be nanoscale actuators and sensors suitable for robots as well as artificial muscles.

It is probable that there will be some limitations to the nanodevices. Memory will be limited and energy carrying capacity will be limited as well. The devices will be able to respond to short messages.

It is also fair to assume that many of the major applications of nanotechnology will be to the improvement of human health and the environment.

The second assumption is that the governmental and regulatory environments will permit nanotechnology to develop with a minimum of hindrances. This may require an expedited patenting process, improved intellectual property laws and adequate insurance mechanisms. The current patent application process hinders development of rapidly advancing technologies. For example, an inventor may have an excellent idea, but before he can bring it into production he will seek venture capital funds. The venture capitalist will insist, not surprisingly, that the idea be patented before investing.

Because the patent application process presently requires a long process, enough time may pass that the inventor's concept is obsolete even before it comes into production. Adequate protection of intellectual property will be essential to the success of nanotechnology. Because nanoproducts will be quite inexpensive to produce, traditional notions of profit margins of up to 30% may be insufficient to justify investment. Further, since nanofactories are quite portable, theft of technology will likely be a deterrent to development. Strong intellectual property laws will insure that investors and inventors will benefit from their work.

The third assumption is that the public will accept nanotechnology. This is not a certainty as past experiences with technological improvements such as nuclear power or genetically modified food have shown. Nanotechnology has an advantage over these technologies however, in that nanotechnology will appear incrementally and will demonstrate obvious consumer benefits. In addition, researchers now understand that the public must have an input to the development of new technologies; perhaps the degree of public input should be guaranteed by law or public policy.

These assumptions are not farfetched. It is already possible to design proteins and enzymes; it will soon be possible to design proteins for specific purposes.

There are substantial costs involved in reaching this point in nanotechnology; however precedent exists for such expenses in such projects as the development of nuclear weapons or the Apollo moon landing effort.

Whether or not the pubic is willing to accept these costs remains to be seen. Both the Manhattan Project and the Apollo Moon Project had a significant impetus; for the Manhattan Project the impetus was propelled by a desire for decisive victory in the war; the Apollo Project was driven by the fear that the United States had fallen behind in the space race. Perhaps the impetus for the development of nanotechnology will be the diminution of sources of fossil fuels. Based on these assumptions, a number of outcomes are likely.

Nanotechnology will encourage the adoption of sustainable processes in industrial production; most of the world will enjoy a First World living standard. By employing highly specific catalysts and focused processing, nanotechnology will allow the production of products on an as needed basis. Because the fabrication can be widely distributed, products can be made at or near the point of use eliminating much of the transportation and storage infrastructure and associated costs. The use of molecular construction methods will blur the line between waste and resource; molecules can be used again and again obviating the need for mines and refineries.

The way in which a corporation derives profit will change. At present, a business creates or conceptualizes a product, gathers the raw material and manufactures the product via a labour intensive production process. At each step, and there are many, the value of the raw material is enhanced and the ultimate cost of the end product increases until the item is ready for retail sale.

Nanotechnology will change this process. The initial design process will be quite complicated and require extensive computing resources to derive the instructions for the nanomachines. Once the machines are designed and programmed, however, the process of making products will be quite automatic and fairly inexpensive (one estimate says that most manufactured goods will be sold by the pound). The present production methods for software provide a model of the process described. For example, the development process for Microsoft Word was long and expensive making the first copy of the program quite expensive, but the second copy of the program cost very little to produce. Because nanotechnology can produce subsequent copies of a product very cheaply, the value and importance of the intellectual property behind the original design will be dramatically increased.

This change in the intrinsic value of products will lead to increased friction between corporations and consumer in much the same way that the controversies surrounding the Napster case caused friction. In a similar fashion there will likely be frictions as traditional factory based manufacturers combine with their labour force to resist nanotechnology based industries. Clean water will be widely available to the over 1 billion humans who do not have access to this resource. Filters based on nanotechnology will allow the cheap and on site production of potable water in vast quantities; even seawater will be treated made potable through low pressure nano based filter materials. As a result, disputes over water rights should diminish.

Information technology will be very widely available. As memory devices benefit from nanotechnology, the cost of memory will decrease and the size of devices will decrease dramatically. As a result, advanced information processing will become widely available. The access to information processing technology will help to enable third world nations to compete in the global marketplace.

Medical advances will allow the treatment of most infectious diseases and many more surgical conditions. In the short term, nanotechnology will provide improved drug delivery systems, for example, medications can be directly targeted to tumor locations. In the advanced state, nanodevices will be able to remove plaque from coronary arteries and other blood vessels thereby alleviating the leading cause of cardiovascular disease. Global energy needs will be met, likely with some variant of solar energy provided at the point of use. With fuel cells using nanotechnology, it will be possible to provide energy via chemical processes minimizing thermal pollution and with much less air and noise pollution.

Off Earth resources will become accessible. The extraordinary strength and light weight of nanomaterials will allow dramatic decreases in vehicle and payload mass making near Earth explorations far less expensive. The notion of a space elevator rests on the development of nanofibre arrays capable of guiding satellites to a low earth orbit. These outcomes, as described, will have important implications for society. The notion of resources will change. For example, because an atom can be recycled again and again, it will not be necessary to remove new raw materials from the soil; this property of nanotechnology will allow enormous strides toward the achievement of a sustainable economy. Areas with abundant strategic minerals, such as copper, tungsten or titanium, will no longer have an advantage over resource poor areas. In a literal and figurative sense, dirt will be as valuable as gold and regions now thought of as desolate will be on a par with resource rich areas.

We will find ourselves in a far more competitive world. National wealth will be less a function of location and conventional natural resources and more a reflection of the technological sophistication of an educated population. In essence, the economic playing field will be dramatically and suddenly leveled and many current national advantages will be zeroed out. Since the means of production can be widely distributed, the traditional advantages conveyed by geography will be minimized. The notion that areas such as the Ruhr Valley are industrial dynamos will be altered. Though access to capital markets will still be important, especially in the early stages of molecular manufacturing, the degree to which a nation displays initiative and will accept innovation will become the primary determinant of the nation's rank among nations.

The size of the population will stabilize or even shrink. As nations have moved from an agrarian to an industrial economy, population growth has stabilized, or in some areas, even reversed. The shift to a nanotechnology based economy will likely continue the trend. As a result, smaller populations may lead to smaller cities with populations susceptible to control measures through the use of nanotechnology based monitoring devices. There may even be competing notions of governance; for example, because nanotechnology will probably lead to a system of large industrial monopolies, population centres may be governed by for-profit corporate entities. Because molecular nanotechnology will decrease the number of manufacturing, transportation and warehousing jobs, life will be less labour intensive. This may, however, lead to fewer wage earners and thus fewer consumers. The structures of taxation may change and lead to novel methods of government finance.

Indeed, the very concept and role of government may be forced to change. In nations with an established history of popular government, the governing bodies may be able to respond to the changes in the industrial base with legislation designed to protect both individual rights and intellectual property. Because the technology will advance quickly, government must seek to avoid tensions and to shape anticipatory solutions. This will require attention to issues such as privacy and new forms of leisure. Significant support must be provided to the educational system to prepare the population for the new demands of a nanotechnology based economy. Because nanotechnology may provide dramatic economic benefits, the government must provide a mechanism to prevent abuse of wealth by industrial oligarchs. It is likely that displaced or unemployed workers will need an extensive social support system as well. The protection of human rights is one of the most important functions of a government and nanotechnology has serious implications for the preservation of human rights. On the one hand, nanotechnology will help provide a safer society though the use of advanced monitoring and tracking technologies, but on the other hand, such techniques may have adverse consequences for the privacy of individuals. One of the major challenges confronting government will be the mediation between the right to privacy and the availability of sophisticated means of surveillance.

As nanotechnology narrows the differences between have and have not nations, ideology will assume a greater role as a source of cultural pride; depending on the nature of the ideology, extremist views may become more prominent to the point of becoming a nidus of conflict. For example, in Western nations, the traditional notion of personal responsibility leads to an internal locus of control, however, Islamic nations, in which a strong tradition of an external locus of control exists (e.g., the will of Allah), it is likely that the rules imposed on man by Allah's will persist despite any benefit from nanotechnology. Nevertheless, the changes to the industrial base from nanotechnology will enable Islamic nations to compete with Western nations on a more level playing field. Such competition will highlight the competitions between ideologies as compared to the competitions between economies.

There are many issues nanotechnology will not mitigate. Regardless of the widespread use of nanotechnology and benefits it offers, notions of racial superiority will not be eliminated. Indeed, the leveling nature of the technology may accent the differences imposed by race or religion. Post World War I Germany had homogenous linguistic, cultural and economic conditions that transcended the lines drawn by religion, but the results of a notion of racial superiority are well known. It is not likely that nanotechnology will erase these fissures in society.

Similarly, nanotechnology probably will not eliminate territorial envy. Nations may harbor desires for adjacent territories based, for example, on historical events such as the Serbian desire to annex Kosovo long after the Battle of Kosovo or the Russian quest for warm water ports which can lead to hostilities even centuries later and despite the benefits brought by nanotechnology. Other sources of friction with the potential to lead to conflict include traditional transnational rivalries such as between the Koreans and Japanese or the Vietnamese and the Chinese. Even within the political borders of nations, nanotechnology will not eliminate ethnic conflicts as between the Hutus and Tutsis.

If nanotechnology results in a decreased interdependence between nations, the threshold for conflict may be lowered. Likewise, because production costs will decrease and products will be widely available, investors will want to insure adequate markets for products. This need for markets may provide an impetus for investor nations to seek to develop captive, or at least dependent, nations as markets. In the American colonial era, a similar predatory arrangement, dubbed Triangular Trade, helped contribute to the American Revolution; Great Britain and China became involved in the Opium Wars for a similar reason. Nations may find it necessary to use force to limit the spread or dissemination of dangerous or proprietary technologies; this may lead to preemptive conflicts. In any case, it is unlikely that the technological advances offered by nanotechnology will abate human self interest or ambition.

Nanotechnology may satisfy many human needs, but it is unlikely that it will satisfy all human wants. No matter how much food or how many possessions people have, they always seem to want more. Thus there will continue to be a drive to create newer, better, faster and more shiny things; thus a military force will continue to be necessary and may become more important.

The Military Impact of Nanotechnology

The impact of nanotechnology extends from its medical, ethical, mental, legal and environmental applications, to fields such as engineering, biology, chemistry, computing, materials science, military applications, and communications.

The potential impact on the environment raises exceptional concerns, as nanoparticles may constitute whole new classes of pollutants. As long as nanoparticles are embedded in common materials (as most of them are in contemporary nanoindustry), they are unlikely to pose any particular environmental threat.

However, as NT advances in the future, larger quantities of pure nanomaterials could be produced. Risks involve the uncontrolled release of nanoparticles, whose potential effect on ecosystems, the environment and the food chain remains largely unexplored and untested. The most drastic scenario of possible adverse effects on the environment - the grey goo scenario - was introduced by Drexler, and it maintains that molecular NT would one day become a reality, and most of manufacturing would be done by tiny nanorobots. These robots would also have to be self-replicating, and Drexler warned that their replication may get out of control.

By consuming all organic material, these autonomous nanorobots could rapidly convert the whole global natural environment into replicas of themselves, the "grey goo", consisting of the "nanomass". Such a scenario is also termed the "global ecophagy". Many scientists are extremely sceptical about the plausibility of such a scenario, attributing it to mere science fiction. As yet, science does not know how to build a self-replicating robot. It is not clear, for example, how those robots would obtain the energy to survive. However, following the precautionary principle, one should not disregard concerns regarding nanorobots as long as they have not been refuted for fundamental or technical reasons.

Nanotechnology, i.e., the science of designing microscopic structures in which the materials and their relations are machined and controlled atom-by-atom, holds the promise of numerous applications. Lying at the crossroads of engineering, physics, chemistry, and biology, nanotechnology may have considerable impact in all areas of science and technology. However, it is certain that the most significant near term applications of nanotechnology will be in the military domain. In fact, it is under the names of 'micromechanical engineering' and 'microelectromechanical systems' (MEMS) that the field of nanotechnology was born a few decades ago-in nuclear weapons laboratories.

A primary impetus for creating these systems was the need for extremely rugged and safe arming and triggering mechanisms for nuclear weapons such as atomic artillery shells. In such warheads, the nuclear explosive and its trigger undergo extreme acceleration (10,000 times greater than gravity when the munition is delivered by a heavy gun).

A general design technique is then to make the trigger's crucial components as small as possible.5 For similar reasons of extreme safety, reliability, and resistance to external factors, the detonators and the various locking mechanisms of nuclear weapons were increasingly designed as more and more sophisticated microelectromechanical systems. Consequently, nuclear weapons laboratories such as the Sandia National Laboratory in the US are leading the world in translating the most advanced concepts of MEMS engineering into practice.

A second historical impetus for MEMS and nanotechnology, one which is also over thirty years old, is the still ongoing drive towards miniaturisation of nuclear weapons and the related quest for very-low yield nuclear explosives which could also be used as a source of nuclear energy in the form of controlled microexplosions. Such explosions (with yields in the range of a few kilograms to a few tons of high-explosive equivalent) would in principle be contained-but they could just as well be used in weapons if suitable compact triggers are developed. In this line of research, it was soon discovered that it is easier to design a micro-fusion than a micro-fission explosive (which has the further advantage of producing much less radioactive fallout than a micro-fission device of the same yield).

Since that time, enormous progress has been made, and the research on these micro-fusion bombs has now become the main advanced weapons research activity of the nuclear weapons laboratories, using gigantic tools such as the US National Ignition Facility (NIF) and France's Laser Mégajoule. The tiny pellets used in these experiments, containing the thermonuclear fuel to be exploded, are certainly the most delicate and sophisticated nano-engineered devices in existence.

A third major impetus for nanotechnology is the growing demand for better materials (and parts made of them) with extremely well characterised specifications. These can be new materials such as improved insulators which will increase the storage capacity of capacitors used in detonators, nano-engineered high-explosives for advanced weaponry, etc. But they can also be conventional materials of extreme purity, or nano-engineered components of extreme precision.

For instance, to meet NIF specifications, the 2-mm-diameter fuel pellets must not be more than 1 micrometre out of round; that is, the radius to the outer surface can vary by no more than 1 micrometre (out of 1,000) as one moves across the surface. Moreover, the walls of these pellets consist of layers whose thicknesses are measured in fractions of micrometres, and surface-smoothnesses in tens of nanometres; thus, these specifications can be given in units of 1,000 or 100 atoms, so that even minute defects have to be absent for the pellets to implode symmetrically when illuminated by the lasers. The final major impetus for MEMS and nanotechnology, which has the greatest overlap with non-military needs, is their promise of new high-performance sensors, transducers, actuators, and electronic components. The development of this field of applications is expected to replicate that of the micro-electronic industry, which was also originally driven by military needs, and which provides the reference for forecasting a nano-industrial boom and a financial bonanza. There are, however, two major differences.

First, electronic devices which can be manufactured in large quantities and at low cost are essentially planar, while MEMS are three-dimensional devices which may include moving parts. Second, the need for MEMS outside professional circles (medical, scientific, police, military) is quite limited, so that the market might not be as wide as expected. For example, the detection and identification of chemical or biological weapon threats through specificity of molecular response may lead to all sorts of medical applications, but only to few consumer goods.

Near and Long-Term Applications and Implications of Nanotechnology

Considering that nanotechnology is already an integral part of the development of modern weapons, it is important to realise that its immediate potential to improve existing weapons (either conventional or nuclear), and its short-term potential to create new weapons (either conventional or nuclear), are more than sufficient to require the immediate attention of diplomats and arms controllers.

In this perspective, the potential long-term applications of nanotechnology (and their foreseeable social and political implications) should neither be downplayed nor overemphasised. Indeed, there are potential applications such as self-replicating nano-robots ('nanobots') which may never prove to be feasible because of fundamental physical or technical obstacles. But this impossibility would not mean that the somewhat larger micro-robots of the type that are seriously considered in military laboratories could never become a reality.

In light of these extant and potential dangers and risks, every effort should be made not to repeat the error of the arms-control community with regard to missile defence. For over thirty years, that community acted on the premise that a ballistic missile defence system will never be built because it will never be sufficiently effective-only to be faced with a concerted attempt to construct such a system! If some treaty is contemplated in order to control or prohibit the development of nanotechnology, it should be drafted in such a way that all reasonable long-term applications are covered.

Moreover, it should not be forgotten that while nanotechnology mostly emphasises the *spatial* extension of matter at the scale of the nanometre (the size of a few atoms), the *time* dimension of mechanical engineering has recently reached its ultimate limit at the scale of the femtosecond (the time taken by an electron to circle an atom). It has thus become possible to generate bursts of energy in suitably packaged pulses in space and time that have critical applications in nanotechnology, and to focus pulses of particle or laser beams with extremely short durations on a few micrometre down to a few nanometre sized targets. The invention of the 'superlaser', which enabled such a feat and provided a factor of one million increase in the instantaneous power of tabletop lasers, is possibly the most significant recent advance in military technology. This increase is of the same magnitude as the factor of one million difference in energy density between chemical and nuclear energy.

In the present paper, the long-term impact of nanotechnology will not be further discussed. The objective is to emphasise the near-to mid-term applications to existing and new types of nuclear weapons.

Nanotechnological Improvement of Existing Types of Nuclear Weapons

Nuclear weapon technology is characterised by two sharply contrasting demands. On the one hand, the nuclear package containing the fission and fusion materials is relatively simple and forgiving, i.e. rather more sophisticated than complicated. On the other hand, the many ancillary components required for arming the weapon, triggering the high-explosives, and initiating the neutron chain-reaction, are much more complicated. Moreover, the problems related to maintaining political control over the use of nuclear weapons, i.e. the operation of permissive action links (PALs), necessitated the development of protection systems that are meant to remain active all the way to the target, meaning that all these ancillary components and systems are submitted to very stringent requirements for security, safety, and reliable performance under severe conditions.

The general solution to these problems is to favour the use of hybrid combinations of mechanical and electronic systems, which have the advantage of dramatically reducing the probability of common mode failures and decreasing sensitivity to external factors. It is this search for the maximisation of reliability and ruggedness which is driving the development and application of nanotechnology and MEMS engineering in nuclear weapons science.

To give an important example: modern nuclear weapons use insensitive high-explosives (IHE) which can only be detonated by means of a small charge of sensitive high-explosive that is held out of alignment from the main charge of IHE. Only once the warhead is armed does a MEMS bring the detonator into position with the main charge. Since the insensitive high-explosive in a nuclear weapon is usually broken down into many separate parts that are triggered by individual detonators, the use of MEMS-based detonators incorporating individual locking mechanisms are an important ingredient ensuring the use-control and one-point safety of such weapons.

Further improvements on existing nuclear weapons are stemming from the application of nanotechnology to materials engineering. New capacitors, new radiation-resistant integrated circuits, new composite materials capable to withstand high temperatures and accelerations, etc., will enable a further level of miniaturisation and a corresponding enhancement of safety and usability of nuclear weapons. Consequently, the military utility and the possibility of forward deployment, as well as the potentiality for new missions, will be increased.

Consider the concept of a "low-yield" earth penetrating warhead. The military appeal of such a weapon derives from the inherent difficulty of destroying underground targets. Only about 15 % of the energy from a surface explosion is coupled (transferred) into the ground, while shock waves are quickly attenuated when travelling through the ground. Even a few megatons surface burst will not be able to destroy a buried target at a depth or distance more than 100-200 metres away from ground zero. A radical alternative, therefore, is to design a warhead which would detonate after penetrating the ground by a few tens of metres or more. Since a free-falling or rocket-driven missile will not penetrate the surface by more than about ten metres, some kind of active penetration mechanism is required. This implies that the nuclear package and its ancillary components will have to survive extreme conditions of stress until the warhead is detonated.

Fourth-Generation Nuclear Weapons

First-and second-generation nuclear weapons are atomic and hydrogen bombs developed during the 1940s and 1950s, while third-generation weapons comprise a number of concepts developed between the 1960s and 1980s, e.g. the neutron bomb, which never found a permanent place in the military arsenals. Fourth-generation nuclear weapons are new types of nuclear explosives that can be developed in full compliance with the Comprehensive Test Ban Treaty (CTBT) using inertial confinement fusion (ICF) facilities such as the NIF in the US, and other advanced technologies which are under active development in all the major nuclear-weapon states-and in major industrial powers such as Germany and Japan.

In a nutshell, the defining technical characteristic of fourth-generation nuclear weapons is the triggering-by some advanced technology such as a superlaser, magnetic compression, antimatter, etc.-of a relatively small thermonuclear explosion in which a deuterium-tritium mixture is burnt in a device whose weight and size are not much larger than a few kilograms and litres. Since the yield of these warheads could go from a fraction of a ton to many tens of tons of high-explosive equivalent, their delivery by precision-guided munitions or other means will dramatically increase the fire-power of those who possess them-without crossing the threshold of using kiloton-to-megaton nuclear weapons, and therefore without breaking the taboo against the first-use of weapons of mass destruction.

Moreover, since these new weapons will use no (or very little) fissionable materials, they will produce virtually no radioactive fallout. Their proponents will define them as "clean" nuclear weapons-and possibly draw a parallel between their battlefield use and the consequences of the expenditure of depleted uranium ammunition.

In practice, since the controlled release of thermonuclear energy in the form of laboratory scale explosions (i.e., equivalent to a few kilograms of high-explosives) at ICF facilities like NIF is likely to succeed in the next 10 to 15 years, the main arms control question is how to prevent this know-how being used to manufacture fourth-generation nuclear weapons. As we have already seen, nanotechnology and micromechanical engineering are integral parts of ICF pellet construction. But this is also the case with ICF drivers and diagnostic devices, and even more so with all the hardware that will have to be miniaturised and 'ruggedised' to the extreme in order to produce a compact, robust, and cost-effective weapon.

A thorough discussion of the potential of nanotechnology and microelectromechanical engineering in relation to the emergence of fourth-generation nuclear weapons is therefore of the utmost importance. It is likely that this discussion will be difficult, not just because of secrecy and other restrictions, but mainly because the military usefulness and usability of these weapons is likely to remain very high as long as precision-guided delivery systems dominate the battlefield.

It is therefore important to realise that the technological hurdles that have to be overcome in order for laboratory scale thermonuclear explosions to be turned into weapons may be the only remaining significant barrier against the introduction and proliferation of fourth-generation nuclear weapons. For this reason alone-and there are many others, beyond the scope of this paper-very serious consideration should be given to the possibility of promoting an 'Inner Space Treaty' to prohibit the military development and application of nanotechnological devices and techniques.

Military Ponders Future of Nanotech

While nano-sciences offer an array of potentially useful technology for the Defence Department, not all military researchers have jumped on this bandwagon.

The disparity of viewpoints was on clear display at a recent lightweight materials conference, where researchers gave differing assessments of its impact.

"If you think biotech has been moving these last few years, just wait," said James Murday, chief scientist for the laboratories at the Office of Naval Research. "The electronics industry will be all nano by the end of the decade."

Other researchers contend the impact will be far less than the predictions of true believers such as Murday. "I'd be shocked if [nanotechnology] will be the thing in 10 years," noted David Stepp, chief of the materials science division of the Army Research Office. "There's always something after."

The federal government is certainly concentrating on nanotechnology. The Clinton administration's 2001 budget raised an informal "nano-scale science and technology working group" to the level of a federal initiative. The National Nanotechnology Initiative (NNI) was born, an effort that has received steady increased in research and development funding.

Federal funding for nanotechnology research and development has increased from \$116 million in 1997 to an estimated \$961 million in 2004. President Bush's 2005 budget request calls for an additional 2 percent increase, boosting the NNI budget to \$982 million.

The Defence Department is one key player in NNI, with a firm interest in developing new materials and manufacturing processes that could determine the shape of a 22nd century fighting force. The Pentagon recognises 10 broad applications of nano-science, including ultra-small computers, low-power communications systems, bio-chemical defence suites, new targeting lasers and countermeasures and innovative materials for armor systems.

Future plans, Murday said, could include further advances, such as directed self-assembly and repair, nano-porous materials that could screen the passage of particles through a membrane and transparent ceramics and composites that could shame current efforts.

Some long-standing engineering problems may find solutions in nanotech. For example, an artificial sapphire used on the nose cones of infrared-guided missiles could be enhanced using advances in nano-science, said Larry Kabacoff, program manager for nano-structured materials for the Office of Naval Research.

If a new material can be formed, it could be an answer to a problem vexing researchers who are searching for a material tough enough to survive insect and weather damage in flight and transparent enough to allow infrared waves through without refracting them. The current systems rely on sapphires which often are damaged while in flight, Kabacoff told National Defence, and using damage-resistant diamonds is cost prohibitive.

Kabacoff's group found that a single-phase material with a structure 20 times smaller than the infrared wavelengths would allow the light to pass unaffected. The program was cut in 2003, "just as progress was being made." A small project with Raytheon still exists, which is generating data that could be used to mount an argument to resume ONR funding. Other nations also are turning their eyes towards nanotechnology, Murday noted. He called China "a major player" who is investing steadily in the emerging technology, and added that Japan is matching the United States' investment dollar-for-dollar.

"We'd just as soon be on the front end of that curve," Murday told the conference attendees.

Social, Military and Ethical Issues in Nanotechnology

Nanotechnology has become a major area of intensive research funding worldwide. The FP6 of the European Commission has set aside 1.3 billion Euro for research within this field in the period 2003-2006. Many countries-with the USA in a leading position-are strengthening their research in nanotechnology similarly. Norway has followed this international trend in science policy through a strategic cooperation between UiO, IFE, SINTEF and NTNU called FUNMAT. An annual budget of 150 million NOK is set aside for ten years in order to increase national expertise in nanotechnology. However, some critical voices like e.g. the etc-group claim that our current knowledge is too limited to handle the risks that one may have to face.

But not only the risks are debated, also the ethics of some possible applications that may be generated are discussed. Nanotechnology travels in miniscule dimensions: one nanometre is one billionth of a metre, while e.g. a human hair is about 80,000 nanometre. Within such a scale one is in effect already below the levels of life and touching the very elements of nature. While many scientists recognise the benefits that the nanotechnology revolution could bring, some have serious concerns. This apprehension is not restricted to one field, but covers human health, environmental impacts, effects on international trade and developing countries, and the possible proliferation in armaments.

A major concern is however that we just do not know what the impacts of nanotechnology will be. It is this lack of knowledge that leads some scientists to call for a moratorium on certain aspects of nanotechnology use and research. The question one needs to confront is whether there is an urgent need for a strict and harmonized regulation of research in this field and how ethical issues related to possible applications could be explored before facts are created that may be difficult to contain or control.

The organisers have invited three international experts that shall address some of the problematic issues in their presentations. Each expert is given ca 30 minutes for presentation and there will be some time for discussion after the lectures. The experts are:

Dr. Jürgen Altmann, Experimental Physics III at the University of Dortmund, Germany, chair of the group "Physics and disarmament", and chair of the "Bochum Verfication-project". Dr Altmann is author of the book Military Uses of Microsystem Technologies: Dangers and Preventive Arms Control, Münster 2001. He has spoken extensively on the possible military uses of nanotechnology in various fora, among others for the Royal Society, London, working group on nanotechnology and nanoscience. He has just finished a 15-months project on preventive arms control for NT (funded by the new German Foundation for Peace Research DSF), however the book will only appear in 2-3 months. An overview can be gained from an article co-authored with Mark Gubrud (Univ. of Maryland, USA). Dr Altmann has made specific proposals for precautionary action in nanotechnology development.

Prof. Vivian Weil is Director of the Centre for the Study of Ethics in the Professions at the Illinois Instiute of Technology, Chicago. Dr. Weil is a Fellow of the American Association for the Advancement of Science, a member of the Governing Board of the National Institute for Engineering Ethics, and Chair of the Executive Committee of the Association for Practical and Professional Ethics. She serves on the editorial boards of various journals, including Science and Engineering Ethics and Science Communication. She received her A.B. and M.A. from the University of Chicago and her Ph.D. in philosophy from the University of Illinois, Chicago. In 1990-1991, she directed the Ethics and Values Studies program of the National Science Foundation (NSF). Recently, she has been involved in NSF conferences to integrate ethics and societal implications in nano science and technology, as a speaker on ethics and as a member of the organising committee of its most recent conference in December, 2003.

Dr. Christoph Baumgartner, from the Interdepart-mental Centre for Ethics in the Sciences and Humanities (IZEW) of the University of Tübingen, Germany. Dr. Baumgartner, born 1969, studied Chemistry and Theology in Tübingen (Germany) and acts as Academic Coordinator of the IZEW. His doctoral Thesis was about "Problems of Motivation in the context of environmental behaviour as a challenge for environmental ethics." His main areas of research are: Environmental Ethics, Problems of Motivation to Morally Responsible Behaviour, Ethical Aspects of Biopatenting, Ethical Aspects of New

Technologies, especially Nanotechnology. He has especially worked with the ethical challenges in medical nanotechnology.

Military Implications of Nanotechnology

In our last issue Mark Gubrud of the Federation of American Scientists started off what we plan to be a long-term debate on potential military uses of nanotechnology. Here Scott Pace of RAND presents another view; it is written from the U.S. perspective, but his points generalize to other nations using nanotechnology.

The Foresight Institute has raised awareness of the many potential social, economic, and political impacts of nanotechnology. The potential military significance of nanotechnology has not yet been examined to comparable depths. The purpose of this article is to raise questions for further discussion of the military and arms control implications of nanotechnology, rather than to offer any definitive answers.

Military capability is but one element of national power addressed by U.S. national security strategy. Other elements are international political influence, economic strength, and national well-being. U.S. national security strategy reflects national interests and consists of a broad plan for achieving objectives supporting those interests. Among the most important objectives are to: "maintain the security of our nation and our allies. The United States, in cooperation with its allies, must seek to deter any aggression that could threaten that security and, should deterrence fail, must be prepared to repel or defeat any military attack and end the conflict on terms favourable to the United States, its interests, and its allies."

Military analysts distinguish between confrontations at peacetime or crisis levels, low-intensity (*e.g.*, guerrilla) warfare, conventional theater war (*e.g.*, the 1967 Arab-Israeli War), theater nuclear war (*e.g.*, use of tactical nuclear weapons in Europe), and strategic nuclear war (*e.g.*, a massive U.S./Soviet nuclear exchange). The United States has a policy of "flexible response" to potential threats across this spectrum of conflict. Different military responses are thus deemed appropriate at different levels of conflict.

How might nanotechnology contribute to U.S. military power at these different levels of conflict? In peacetime or crisis, nanocomputers may allow more capable surveillance of potential aggressors. The flood of data from worldwide sensors could be culled more efficiently to look for truly threatening activities. In low-intensity warfare, intelligent sensors and barrier systems could isolate or channel guerrilla movements depending on the local terrain. In conventional theater war, nanotechnology may lead to small, cheap, highly lethal anti-tank weapons. Such weapons could allow relatively small numbers of infantry to defeat assaults by large armored forces.

At nuclear conflict levels, accurate nanocomputer guidance and low nanomachine production costs would accelerate current trends in the proliferation of "smart munitions." Rather than requiring nuclear weapons to attack massive conventional forces or distant, hard targets, nanotechnology enhancements to cruise missiles and ballistic missiles could allow them to destroy their targets with conventional explosives. Conventional explosives themselves might be replaced by molecular disassemblers that would be rapidly effective, but with less unintended destruction to surrounding buildings and populations. President Reagan's goal of making nuclear weapons "impotent and obsolete" could be reached not by space-based defences, but by terrestrial nanoweapons making nuclear weapons irrelevant.

These potential military applications would allow the United States a greater range of options in deciding how to respond to aggression. Nanoweapons could lower the cost of meeting aggression (in both dollars and lives) in tactical applications while preserving strategic deterrence without nuclear weapons. Strategic deterrence today is not limited to deterring nuclear attacks on the U.S. homeland, but on preventing political coercion of the U.S. and its allies. Should our allies be threatened by a third country, the U.S. could either respond directly with its own forces or provide support to our allies, including nanotechnology weapons, to preserve their security. Nanoweapons could deter war either by threatening unacceptable damage to an aggressor, as with today's strategic nuclear weapons, or by denying any plausible achievement of an aggressor's objectives, as is the potential with space-based missile defences.

How might nanotechnology affect the future of nuclear and conventional arms control?

How might nanotechnology affect the future of nuclear and conventional arms control? Arms control has traditionally focused on three objectives in support of U.S. national interests. The first is to lower the possibility of war. The second is to lower damage caused should war occur.

The third is to lower the burden of armaments on society. Nanoweapons may provide new temptations to aggression should one nation or group of nations achieve a great enough technical lead in the field. In achieving the first objective of arms control, the U.S. may seek to either limit knowledge of nanotechnologies (as is done with nuclear weapons) or seek to disseminate knowledge widely to allow development of countermeasures to potential nanoweapons.

In terms of limiting damage should deterrence fail, nanotechnology may allow great increases in target discrimination so that noncombatant casualties and damage are minimized. Conversely, nanotechnology could be used to create weapons of mass destruction, such as chemical, biological, and nuclear weapons. The first nanoweapons of certain classes will raise interesting questions under the international law of armed conflict and treaties proscribing biological weapons and environmental modification techniques.

As for lowering the economic burden of armaments on society, nanomachines may lower the cost of producing weapons and their supporting command and control systems. The United States may choose to use this new economic productivity to create more capable forces, rechannel defence expenditures to other needs, or lower taxes. Perhaps all three outcomes will occur.

U.S. air, naval, and ground forces all have distinct roles and missions to perform that are not likely to disappear with changes in technology, although some of the changes may be dramatic as when the Army's cavalry gave way to the tank. Nanotechnology does not allow one to violate the laws of physics; visible light sensors built with nanomachines will still not be able to see through walls.

The most lethal nanoweapon is not very useful if it cannot find or reach its target. A U.S. armed with nanoweapons may be able to deter or prevent attack on its homeland, but it will continue to have obligations to allies who may ask for help. Discussions of military applications of nanotechnology will likely be most productive if grounded in realistic appraisals of nanotechnology limits, actual military needs, and continuing U.S. national interests.

Military, Arms Control, and Security Aspects of Nanotechnology

The announcement of the US National Nanotechnology Initiative and its counterparts in other countries has been accompanied by expressions of concern about ethical, legal and social implications, and some allotment of funding to address them, but concerns arising from military uses of nanotechnology, presumably in the service of national security, have been largely left out of the working definition of "societal and ethical implications". From the perspective of international security and arms control analysis it appears that systematic study by scholars in the hard and social sciences has hardly begun, with only a few scholarly articles published.

This is curious, since the parallel popular discourse on nanotechnology (NT) has not failed to notice that promises made for the possibilities of NT would in practice have profound implications for military affairs as well as relations between nations and thinking about war and international security. Superweapons made possible by nanotechnology are stock items by now in science fiction, and to some extent that literature has already begun to explore the technical logic to see where NT may lead us.

Military writers have also taken note of these emerging ideas, while in the USA the military enthusiastically spends one quarter to one third of all Federal nanotechnology research and development funds, and its visionary powerpoint artists portray a future of nano-enabled supersoldiers fighting on nanotech battlefields. Here we try to consider the impact of evolving and possibly disruptive military NT applications within the context of history, conflict, international security and world order.

The international community seeks to avoid war through a variety of mechanisms, including military deterrence, international organisations and treaties, arms control and disarmament. These mechanisms are created and evolve in a changing technological environment. Arms control in the past has included restrictions on new or emerging military technologies. Today, emerging military NT applications and their consequences need to be analyzed. We advocate preventive arms control where one can identify specific negative impacts which may be subject to feasible controls.

New Military Technologies

The purpose of armed forces is to break the will of an organised opponent by violent force. Who will prevail depends to some extent on comparative recklessness. Thus, war and the preparations for it do not only go beyond civilized behaviour, but carry an inherent tendency to transcend all other rules, including the ones applying to armed conflict. Innovation in military technology can become an extension of warfare itself. Being ahead in technology provides an important advantage in armed conflict, and if the lead position is not attainable, then one should at least stay as close as possible behind. Thus, all potential opponents have constant strong motives for military research and development and for incorporating their results into the armed forces. Secrecy and worst-case assumptions – both to some extent necessary elements of military preparation – increase these motives.

Military applications of new technologies take place in a framework that is quite different from civilian ones: the bad and ugly uses are not results of accidents or criminal actions – they are prepared in an organised way on a large scale by and with the resources of the state. A special problem arises here: states have to reckon with an opponent using any means at hand. If technology allows new, more effective weapons, they might be used– even if they violate the law of warfare. In order to protect oneself, one needs to know the characteristics of the new weapon; this creates a motive to research and develop the new weapon oneself – which, in turn, creates mistrust and fear between potential opponents.

An example of this mechanism can be seen in the debate regarding biological-weapons – what constitutes legitimate defencive research, and where does forbidden development of biological-warfare agents begin? Of course, the economic capacity of the state limits its pursuit of weapons, and limitation is also possible by political decision. Self-restraint may be exercised in respect of national or international public opinion. Limits are also implied by the international law of warfare. Finally, potential opponents may agree on mutual limitations (arms control). In general, for these to be effective, there need to be reliable ways of verifying compliance.

General Aspects of War and Peace

Throughout history, improved technology has provided advantages in battle. In the 20th century, science and technology became tightly integrated with the system for waging and preparing war. Following the Second World War, military research and development (R&D) continued to expand, in particular in the USA and USSR.

The 1950s to 1970s, roughly, marked the era of "big science" – big in budget and in physical scale, while often in pursuit of control over the very small. Nuclear bombs, nuclear-driven submarines, long-range ballistic missiles, and orbiting satellites mark some of the important milestones in the qualitative and quantitative arms race of the Cold War – achieved at extremely high cost and effort. Even though the Cold War is over, military threats are still at work as instruments of geopolitical will as well as the basic mechanisms of national and international security. Nuclear deterrence is a doctrine still in effect, and thus nuclear war could still start at any time, although this may be improbable. Despite massive reductions (from a high of 65,000 in 1986), there remain about 20,000 nuclear warheads on Earth. The experience of the World Wars has led to a fundamental change in international law.

Through the centuries, states assumed a natural right to go to war – for whatever reason or purpose. With the founding of the UN and the acceptance of its Charter in 1945, maintaining international peace and security became the central imperative (Art. 1). The use of force and threat to use force in international relations were forbidden (Art. 2), with few exceptions: one is force legitimized by the UN Security Council to restore peace and security (Chapter VII, Art. 39-51), and the other is in self-defence until the Security Council has taken its measures (Art. 51).

The security mechanisms foreseen in the UN Charter were hardly implemented, and nuclear war after 1945 was not prevented by adherence to its articles, but rather by the threat of mutual annihilation. Yet the norm established in the UN Charter has had strong effects on the international community. Wars waged since 1945 have usually been claimed to be defencive, and when such claims have been rejected the offending power has often come under collective pressure to cease or restrain its aggression. Of course, the principle of refraining from the use of force except in self-defence is severely endangered if preventive wars are being waged based on perceptions of threats, not on actual aggression.

Thus, we believe the 2003 US war against Iraq has weakened the UN and the international mechanisms of maintaining and restoring peace and security. When armed conflict occurs, regardless of whether it constitutes legitimate or illegitimate use of force according to the UN Charter, the warring parties are limited in their choice of means and methods of warfare by international humanitarian law, which consists of basic as well as specific rules, often written down in international agreements, which evolve over time.

Many rules have become customary international law and are binding for all parties to a conflict; this holds, *e.g.*, for the principle of attacking only military targets or of humane treatment of combatants *hors de combat*. Some rules are only obligatory for the respective signatory parties, *e.g.*, the Additional Protocol 2 to the Geneva Conventions of 1949 on non-international armed conflict. Before introducing a new weapon, means or method of warfare, the states are obliged to check whether its employment would be prohibited by international law.

Arms Control and Disarmament

Despite the nuclear threat that loomed throughout the 1950s, limitation talks between the Cold War antagonists were unsuccessful until the experience of the Cuban missile crisis (1962) when nuclear war seemed imminent. Real progress in arms control started with the Partial Test Ban Treaty in 1963.

Important multilateral treaties that followed included the Outer Space Treaty (1967) and the Nuclear Non-Proliferation Treaty (1968). Political relations between the USA and the USSR improved enough to permit the bilateral Anti-Ballistic Missile (ABM) Treaty and SALT I (1972), later SALT II (1979). Another tense period of confrontation and impasse in bilateral arms control followed, until the Soviet approach changed under Gorbachev, leading to the INF and START I Treaties (1987/1991).

Full disarmament, that is reduction to zero, was agreed for biological weapons (1972, multilateral), and chemical weapons (1993, multilateral). Nuclear test explosions (1996), and anti-personnel land mines (1997) have also been banned. These agreements have been mostly successful, despite the lack of full participation by all relevant powers, and despite some serious violations which have been detected and exposed – most notably with respect to the Biological Weapons Convention which was implemented without provisions for verification, inviting contempt. General and complete disarmament (all countries, all armaments and armed forces) has been the declared goal of the UN and was mentioned in several arms-control treaties, but this goal has not been seriously pursued in actual policy.

Arms control must be considered an unfinished project, and important gaps remain: there is no ban on nuclear weapons, there is no prohibition on space weapons (except for weapons of mass destruction deployed in space), and limitations on conventional arms and forces exist only in Europe. Recently, arms control and humanitarian law have become endangered by actions of the USA. Even before Sept. 11, 2001 the US refused to ratify the

Comprehensive Test Ban Treaty and blocked the negotiations on a Verification Protocol to the Biological Weapons Convention. Recently there have been moves towards easier re-sumption of nuclear tests. In 2002, the USA abrogated the ABM Treaty. The US has also systematically sought to undermine the International Criminal Court that was instituted by the international community.

Military Nanotechnology

Nanotechnology will change many aspects of our lives. Powerful computers will be ubiquitous. With the advance of artificial intelligence, the replacement of human labour by artificial systems will accelerate. New materials will lead to higher energy efficiency. Therapeutic drugs will be designed for the individual. Along with opportunities, there are also potential risks, be it to health, the environment, social justice or privacy. Convergence of nano-scale, biomedical, information, and cognitive science and technology (NBIC) can lead to applications with profound impacts on the human condition. The 2001/2 U.S. workshop on NBIC convergence mentioned, *e.g.*, nano-implant devices, slowing down or reversing aging, direct brain-machine interfaces, human-like artificial intelligence.

These and other far-reaching concepts of manipulating the human body and mind imply risks and dangers, as well as ethical challenges, on an unprecedented scale. Containing abuse and unintended consequences will be difficult even in the civilian realm. Military NT applications pose special risks – first, because of the reparations for destructive uses and second because of secrecy. Tackling this problem calls for special efforts. In order to provide reliable information, one can look at actual military NT research and development in the USA, which is both the leader in military NT and also much more transparent about its military research and development than any other country. In addition, one can extrapolate scientific-technical advances and assess what military applications will become possible in principle.

Military NT R&D in the USA: In the USA, military research and development in nanotechnology (NT) has surged. Of the funds for the National Nanotechnology Initiative, one quarter to one third goes to the Department of Defence – in 2003, $ 243 million of $ 774 million. This is far more than any other country – the UK expenditures, for example, were stated as about $ 2.6 million in 2001. Assuming total West European funding five times as high, with similar levels respectively for Russia, China and the remaining countries that are active in military NT R&D, the US expenditure would be five times the sum of all the rest of the world. (In military R&D at large, the USA accounts for two thirds of the global total.)

U.S. work spans a wide range in the spectrum from basic research to advanced technology development – development of actual systems for deployment is still several to many years off. University research grants fund nanoscale machines, carbon nanotubes, quantum computing and magnetic nanoparticles. The Defence Advanced Projects Agency funds projects in magnetic memory, bio-computing, bio-molecular motors, sensors for chemical and biological warfare agents, and micro robots, among many others. The research laboratories of the armed services work on self-assembly of nanostructures, organic light-emitting diodes, carbon nanotubes and composites, nanomaterials for explosives and propellants as well as for armor and projectiles, and many similar topics.

In 2002, the Army selected the Massachusetts Institute of Technology to house an Institute for Soldier Nanotechnologies, with up to 150 staff to work in seven multidisciplinary research teams. Their goals include a battle suit that protects against bullets, chemical and biological agents, and stiffens on demand to act as compress or splint. Sensors are to monitor the body status. For carrying heavy loads, an exoskeleton with "muscles" from artificial molecules is envisaged.

Potential Military Applications of NT: In general, NT can lead to improvements in traditional military systems and to qualitatively new ones. Very small but highly capable computers will be used in weapons, uniforms, logistics, and communication systems. Increasingly sophisticated and discriminating sensors may become very small, and cheap enough that they can be scattered in high numbers to saturate an area, ostensibly yielding "total awareness". Guns will shoot farther, projectiles and missiles with cheap guidance systems will become smaller and more accurate. Vehicles will become lighter and more agile, with more powerful engines and greater range. Energy storage is a key problem for many military systems, and NT is often considered a key to solving it.

Autonomous vehicles (robots) for reconnaissance and communication, but also for fighting, will arrive; some of them may be very small. NT ultimately raises the prospect of even microscopic mobile robots, although macroscopic vehicles are needed for high speed or long distance travel. Sophisticated fighting robots, the successor to today's killer drones and prototype robot combat planes (UCAVs), will be enabled by advanced computers, smart materials, advanced energy and propulsion systems, and other NT-based refinements.

Similarly, NT will contribute to lowering the cost and increasing the capability of space systems, including possibly very small antisatellite weapons. Robotics will be used in logistics, production and automation of complex weapons systems. A key enabler of robotics applications will be advanced computers capable of situation assessment and action planning, for example the motion planning needed to coordinate dextrous manipulators, or to maneuver through "battlespace".

New chemical or biological warfare (CBW) agents may become possible that act selectively only against the intended targets. Nanobiotechnology is ambitious enough to propose robotized artificial microbes, which could become tools of assassination or mass murder. At the same time, nanomaterials for filtration and neutralization, and NT-based sensors and nanomedicine in general may provide new approaches to CBW defence. Advances in biocompatible materials and portable biomedical systems may allow the creation of body implants to monitor health status, release drugs or interface to the nerves and brains of fighters. Their portable computers may evolve into wearable information appliances producing an "augmented reality" which simultaneously gives the fighter access to information from the net and also gives command access to the soldier, placing her under some degree of "remote control".

The tendency toward cyborgization follows directly from such military goals and culminates in the vision of direct brain-technology interfaces, but the possibility of improving in this way on the performance of well-trained human senses and bodies, whether for fighting or for piloting or for thinking interactively, seems remote. NT can be used in enhanced versions of existing nuclear weapons incorporating improvements to safety, reliability, etc., or possibly new types of conventional explosive in the fission primary. More speculative concepts include qualitatively new weapon types such as pure-fusion explosives of arbitrarily small yield. It is hard to predict how NT advances may impact nuclear weapons production and barriers to proliferation, but we have seen substantial advances in these technologies since 1945 and further improvement seems possible.

Scenarios along these lines, of more or less visionary character, are being discussed within the military and national security establishments of the world, particularly the USA, but many of these concepts will not prove militarily effective, as has been the case through-out history. Countermeasures to NT weapons will exploit NT as well, giving rise to complicated correlations of forces and complex arsenals within which unexpected interactions can arise.

Preventive Arms Control

Preventive arms control is qualitative arms control applied to the future; it is about stopping or at least limiting dangerous military developments before they become actual, in particular weapons exploiting or based on new technologies. Limitations can be designed to intervene at the stages of development or testing, or sometimes research. Precedents include:

i) the ABM Treaty, which prohibited not only deployment, but also development of anti-ballistic missile systems that are sea-based, air-based, space-based or mobile land-based;

ii) the nuclear testing treaties, which preclude certain experiments with actual nuclear explosions; and iii), the Protocol on Blinding Laser Weapons (1995), which explicitly bans only the use of blinding lasers but led to halting their development.

Preventive arms control assesses a potential new military technology using several criteria. One group of criteria concerns dangers to arms control and the international law of warfare: will the new technology undermine existing controls, law and norms? Another set of criteria concerns stability: will the new technology destabilize the military situation, *e.g.*, by reducing reaction times? Will it lead to a technological arms race? How about proliferation, in particular to regions with high probability of conflict?

Finally, one needs to consider unintended hazards to humans, society or the environment. If there are good arguments for restricting a particular technology, then one needs to take into account any positive uses – in particular in the civilian realm – and if possible devise rules that exclude the negative applications without overly restricting the positive ones. Methods for verifying compliance also have to be thought out – they must provide assurance that illegal activity would be detected, but must not be too intrusive or too burdensome.

Preliminary Assessment

When considering the various potential military NT applications under criteria of preventive arms control, one finds several that will be close to civil uses, such as small, fast, distributed computers or strong, light-weight structural materials. A few uses could help to protect against terrorism or would act mostly defencively. Examples are sensors and decontamination agents for biological weapons or improved injury care. Preventive limitation would be unrealistic or counterproductive in such areas. However, there are several potential military applications of NT and/or NBIC at large that raise serious concerns under criteria of preventive arms control. In the medium term, the most dangerous ones involve:

- New selective chemical or biological warfare agents: These would violate the existing conventions, while posing new challenges to verification; they could be used either as weapons of mass destruction or for targeted assassinations, not only by armed forces but also by terrorists.

- Autonomous fighting systems – robots and robotic vehicles on land, in water or air: They would violate the international law of warfare if they would produce superfluous injury or could not recognise non-combatant or combatants *hors de combat.* Autonomous tanks or combat aircraf considered outside of the definitions of the Treaty on Conventiona Armed Forces in Europe could undermine and endanger that treaty Small satellites capable or attacking other satellites by direct hit or by manipulation after docking would destabilize the situation in outer space – not without consequences on Earth.
- Mini-/micro-sensors and-robots, including biological/artificial hybrids They could be pre-deployed covertly in an opponent's territory to strike at an appointed time or on command, or to guide other weapons. If such devices are produced at low cost in large numbers, diffusion to other countries and to criminals is probable, creating the possibility of their use in asymmetric warfare or terrorist attacks.
- Body manipulation including implants: Under the imperative of combat efficiency, armed forces may more readily explore new possibilities for body manipulation than civilian society. Soldiers might voluntarily, or maybe under some pressure, accept risky or ethically questionable technologies that modify body chemistry, rewire brain, nerve and muscle or otherwise radically alter the human organism. This could create "facts" and circumvent barriers in civilian society, preempting a thorough debate on benefits, risks, ethical aspects and needs for regulation.

Aspects of Molecular Nanotechnology

Particularly dramatic issues are posed by the vision of "molecular" or "machine phase" nanotechnology (MNT) as conceived by Drexler, Merkle and others. In this vision, complex systems would be structured, like living systems, from the nanoscale up. Using tough materials and machinelike principles not found in naturally evolved organisms, they could manufacture products from molecular components in lifelike processes similar to growth and self-replication. The products of such a technology would have performance characteristics far beyond what is achievable today.

For example, these might utilise carbon materials with fully-integrated nanostructure and nanosystems. Artificial intelligence supported by MNT hardware should easily surpass human capabilities and could be used to direct the production and use of MNT systems without requiring human supervision. Realisation of this vision would create extreme dangers under all the criteria of preventive arms control, with the greatest problems arising from the potential for an arms race using autonomous production, and destabilization by pressures to attack first.

The first task in addressing this area is to provide reliable assessments of the feasibility and time frame to develop the technology. MNT proponents have suggested that their vision could be realised in as little as a few decades. If this turns out to be a realistic possibility, the regulation, including international preventive arms control, needs to be developed well in advance.

Topics for Further Research

Concrete interdisciplinary investigations in the coming years should examine military NT programs in several countries, including the USA, potential opponents (NT-capable as well as threshold countries), U.S. partners and friends. This should be done with a view to building cooperation and avoiding exaggerated perceptions as motives for NT arms races. Other important topics for research include the potential for terrorist uses, and the use of NT to improve verification of compliance with agreements.

Concluding Remark

Containing NT dangers in civil society will require rules, checks, and penal measures comparable to those in effect for dangerous chemicals, nuclear technologies, genetic modification of pathogens, etc. Reliably preventing dangerous military uses by international agreement will need similar degrees of monitoring and juridical prosecution. This is difficult to conceive in an international system where national security is built on the threat of using armed forces.

Unfortunately, the logic of today's international system points to confrontations, at the level of deterrence at least, between states armed with nanotechnology-based weapons. As advanced capabilities mature, this could become a new arms race tending toward instability. We better decide early which road we are to take, and it better be the one that leads to regulation.

Will powerful new technologies act as a lever to qualitatively strengthen the rule of law and other elements of civil society in the international system? Or will they provide ways to circumvent existing rules and render them ever less meaningful, ushering in an unregulated future in which technology itself dictates the fate of humanity?

Recommendation for Preventive Arms Control

These dangers can be contained by preventive arms control. Regulation need not focus on NT as such, but should take a wide view and address military mission areas. In many cases, the dangerous NT uses come under the headings of general agreements that exist already or that the international community has long asked for. In concrete terms, we recommend:

- The existing arms control and disarmament treaties that are in force as well as humanitarian international law should be respected and preserved. The Biological Weapons Convention should be augmented by a Verification Protocol.

- A general ban on space weapons should be adopted, with special rules for small satellites.
- There should be limits on military autonomous vehicles and robots, in particular with a combat function. Particularly important is a ban on autonomous killing.
- Small mobile systems should be mostly prohibited in the military as well as in civilian society, with very few exceptions (*e.g.* for search of collapsed buildings).
- Implants and other body manipulations that are not motivated by a direct medical condition should be subject to a renewable moratorium of ten years' duration. For consistency and completeness, all rules for the military need to be coordinated with the regulation that is to be developed for the civilian realm.

Military Reloads with Nanotech

Nanotechnology is grabbing headlines for its potential in advancing the life sciences and computing research, but the Department of Defence (DoD) found another use: a new class of weaponry that uses energy-packed nanometals to create powerful, compact bombs.

With funding from the U.S. government, Sandia National Laboratories, the Los Alamos National Laboratory, and the Lawrence Livermore National Laboratory are researching how to manipulate the flow of energy within and between molecules, a field known as nanoenergentics, which enables building more lethal weapons such as "cave-buster bombs" that have several times the detonation force of conventional bombs such as the "daisy cutter" or MOAB (mother of all bombs).

Researchers can greatly increase the power of weapons by adding materials known as superthermites that combine nanometals such as nanoaluminum with metal oxides such as iron oxide, according to Steven Son, a project leader in the Explosives Science and Technology group at Los Alamos. "The advantage (of using nanometals) is in how fast you can get their energy out," Son says.

Son says that the chemical reactions of superthermites are faster and therefore release greater amounts of energy more rapidly.

"Superthermites can increase the (chemical) reaction time by a thousand times," Son says, resulting in a very rapid reactive wave. Son, who has been working on nanoenergetics for more than three years, says that scientists can engineer nanoaluminum powders with different particle sizes to vary the energy release rates. This enables the material to be used in many applications, including underwater explosive devices, primers for igniting firearms, and as fuel propellants for rockets.

However, researchers aren't permitted to discuss what practical military applications may come from this research.

Nanoaluminum is more chemically reactive because there are more atoms on the surface area than standard aluminum, according to Douglas Carpenter, the chief scientific officer at nanometals company Quantumsphere.

"Standard aluminum covers just one-tenth of one percent of the surface area (with atoms), versus fifty percent for nanoaluminum," Carpenter says.

Carpenter says the U.S. military has developed "cave-buster" bombs using nanoaluminum, and it is also working on missiles and torpedoes that move so quickly that they strike their targets before evasive actions can be taken.

"Nanoaluminum provides ultra high burn rates for propellants that are ten times higher than existing propellants," says Carpenter. The military is also trying to make sure that its bullets kill quickly.

The U.S. Army Environmental Centre began a program in 1997 to develop alternatives to the toxic lead that is used in the hundreds of millions of rounds that are annually fired during conflicts and at its training ranges. Carpenter says that although bullets using nanoaluminum are ready to be field tested, the government has been slow implement the technology. "Getting the government to change the way they kill people is difficult," Carpenter says.

Because nanometal provides a higher concentration of energy while requiring fewer raw materials, the overall cost of these weapons would drop, according to Kevin Walter, vice president of technical business development at nanometals manufacturer Nanoscale Technologies.

"You get a little better bang for your buck," Walter says.

The nanometals can be produced in particles as small as eight nanometres, Walter says, and then combined with other chemicals to create the explosive materials, which can also be used for non-military applications including pyrotechnics and explosives for mining.

Nanotechnology "could completely change the face of weaponry," according to Andy Oppenheimer, a weapons expert with analyst firm and publisher Jane's Information Group. Oppenheimer says nations including the United States, Germany, and Russia are developing "mini-nuke" devices that use nanotechnology to create much smaller nuclear detonators.

Oppenheimer says the devices could fit inside a briefcase and would be powerful enough to destroy a building. Although the devices require nuclear materials, because of their small size "they blur the line with conventional weapons," Oppenheimer says.

The mini-nuke weapons are still in the research phase and may be surreptitiously funded since any form of nuclear proliferation is "politically contentious" because of the possibility that they could fall into the hands of terrorists, Oppenheimer says.

The creation of much smaller nuclear bombs adds new challenges to the effort to limit weapons of mass destruction, according to Oppenheimer.

"(The bombs) could blow open everything that is in place for arms control," Oppenheimer says. "Everything gets more dangerous."

Risks of Military Research in Nanotechnology

Alexis Vlandas is a graduate re-searcher at the University of Oxford working in the field of nanotechnology. The present paper follows his presentation at an INES seminar during the European Social Forum in London, November 2004.

The "Nano" Buzz

Although nanotechnology might not yet be the word on everyone's lips, it certainly has been causing quite a buzz in academic and industrial cir-cles in the past years. This word, often found inadequately defined by its proponents and opponents alike, has been recently quoted in sentences predicting either heaven or hell on Earth. The polarisation of the debate early in the development of this technological field is reminiscent of the Genetically Modified Organisms (GMO) debate of the 90's. To pre-vent a similar backlash, the scientific community and emerging business stakeholders have realised that steps were to be taken as early as possible. Communicating to the public became a priority, with the actors involved facing an increasingly up-hill battle as the public's minds started to be filled with the fictions and fears inspired by nanotechnology.

Among these efforts, came a distinct view largely pushed by NGOs and scientists to seize this opportunity to move away from a "knowledge deficit" communication approach to a more participatory one. In the former, the aim was to educate the public to enable it to have the "right" opinion on an issue. It presupposed that any concerns raised by non-experts are the result of a lack of information. Recognising that this could no longer be the way to address concerns arisen around a scientific issue, a more open and participative process was advo-cated. The nanotechnology report published in March 2004 by the Royal Society illustrates this new approach.

In assessing the risks involved with the development of nanotechnology, several points should be kept in mind. Firstly, one must distinguish nanoscience – the advancement of human understanding of natural phe-nomena – and nanotechnology which is the application (via products) of this science. Secondly, nanotech-nol-ogy is not a unified field. While one could make a case that the GM debate revolved around a single tech-nique, the nano field is extremely diverse encompassing numerous techniques and instruments. It could indeed be argued that we should actually refer to nanotechnologies rather than nanotechnology. Lastly, the risk/benefit assessment should include questions such as: Who does what? In the benefit of whom? Who experiences associated risks and who carries the responsibility?

What's Nanotechnology?

The prefix "nano" means a billionth or 10-9. Nanotechnology and nano-sciences are therefore concerned with devices and phenomena talking place on the nanometre scale (it is often considered that anything under 100 nm qualifies as nano). As a comparison a carbon atom has a radius of roughly 0.1 nanometres. It can be argued that the synthesis of objects in this dimension is hardly new. In fact, the chemistry of small molecules always produces nano-scale entities. Similarly, the wood fires (or car engines) release nano particles. So what is different now? Why are we talking of nanotechnology as the next industrial revolution?

It was early on realised that if one could design a material with atomic resolution, the properties such material would exhibit could be entirely new. This is due to the fact that under a certain size, matter stops behaving according to the rules of classical physics and begins to act according to the ones of quantum physics. By engineering material carefully, one can arrange the atoms so that we can use this different set of rules in a useful way. To achieve such a feat, two distinct areas have to be developed in parallel: the understanding of the physical laws governing matter at this scale and the instrument enabling the manipulation and characterisation of matter at this scale. These two fields "feed" one another, advancement in one enabling further progress in the other.

A crucial step was made in 1981 with the development of the Scanning Tunnelling Microscope or STM. Relying on a quantum mechanical effect, this instrument enables for the first time researchers to image individual atoms resting on a surface. Soon after it was discovered that, using the same apparatus, one could probe the physical properties of an atom or group of atoms. Lastly, it was noticed that you could also use an STM to move individual atoms. This produced the well known STM picture in which the letter "I", "B" and "M" were written with xenon atoms.

The ability to design, synthesise and study objects with at least one of their dimension in the nanometre scale is what can be called nano-technology. The level of control we can attain now gives us the opportunity to create a whole new range of devices: from self cleaning glass, to new catalysts, better batteries, stronger materials or bio-compatible bone substitute. As one can see, it is a very vast field of science and technology where many different techniques come into play. A very interesting corollary that comes with working at this scale is that to a large extent the boundaries between what we know as physics, chemistry and biology are abolished. Indeed one can use semiconductor production techniques to create a series of trench on a material, coat them with hydrophilic functional groups to make them suitable to transport water, add "nano-engine" constituted by ATP converting bio molecules (taken from animal cells) to move the water around and finally insert molecular sensors: you obtain what is called a "lab on a chip."

Nanotechnology will impact fairly fast nearly every domain of the manufacturing and electronics industries. It will then slowly diffuse to the health related industries. It will also most probably yield in its progress a number of important scientific discoveries. There is one trait that might make nanotechnology different from previous scientific and industrial revolutions (at least for a while): its apparent absolute character. Indeed, as there are few prospects to go for the sub-atomic level in common appliances due to the energies that it would involve, nanotechnology is seen by some as a kind of "ultimate frontier" for the control of matter.

Funding and Why is the Military Interested?

Funding for nanotechnology and nanosciences has increased tremendously in the past years across the globe. The scale of the discoveries to be made in this field as well as the potential for commercially viable applications are two of the many reasons behind this increase. It is believed that the bulk of today's effort comes from states with private companies providing much less. This is however hard to check as detailed information on research investment of companies is understandably not easily available.

The EU (member states and Euro-pean Commission), for example, spends roughly 800 millions euros each year on nano-related research and this figure is planned to increase in the subsequent years. In the USA, the National Nanotechnology Initiative (NNI) was launched in 2000 to provide an integrated framework for nanotechnology research, the 2004 fiscal year request for this initiative is of 849 millions dollars. The sum is shared between several federal agencies such as the NSF, the NIH or the DoE. In 2003 however, none of these agencies were the main recipient of NNI funds.

The department of defence (DoD), with 243 millions dollars out of a total of 774 millions, received 31.3% of total allocated resources. In Europe as well, the military is one of the key investors in this new area even if getting the information is harder. Sweden will spend 11 millions euros in the next 5 years on military nanotechnology research programs through the FOI (Swedish defence research agency). In the UK, the Ministry of Defence (MoD) spends roughly 1.5 million pounds/year on nano R&D. This will mostly go toward research on new structural materials and electronic devices. It was decided that the MoD would mostly rely in the short-term on civilian nanotechnology but would establish a technology watch on the subject.

It is worrying to see that such a disproportionately important part of the research effort in this field is funded by the military. What aims is the defence sector pursuing in funding research in this area? What are the associated risks?

The Strategic Vision and Examples of Programs

The vision that drives military research is not one filled with nano-robots. It is not today envisioned that nanotechnology itself will provide new types of weapons. The changes brought by nanotechnology will be of a more pervasive nature. In the words a Clifford Lau, the Pentagon senior science adviser: *"Nanotechnology is a 'force multiplier'. It will make us faster and stronger on the battlefield."*

In the short term, nanotechnology will be used in continuity with previous programs developed by relying on "traditional" material science, microelectronics and chemistry. The improvements in materials, using nanostructured materials, will provide stronger and lighter armour, reduce maintenance costs and marginally improve the performance of kinetic weapons (better penetration of shells, etc). The shrinking of electronic circuits will enable sophisticated communication devices and sensors to be deployed on a near permanent basis (using better energy sources).

The survivability of troops will also be improved by the insertion of sensors in their uniforms providing real time medical monitoring and deployment of chemical and biological agents' sensors. In the medium term, the army will be seeking to improve human performance on the battlefield by means of an exoskeleton. All these technical advances will combine to produce an increase in the soldiers' "lethality." The US army anticipates that the deployment of troops with capabilities so far beyond that of their enemies would provoke awe and thus ensure victory.

The nanosciences and nanotechnology advances will also affect weapons development in the field of bio, chemical and nuclear weapons in several ways. For example, it is believed that the nano-structuring of nuclear material pellets is essential to provoke chain reactions in laser driven nuclear research facilities such as the National Ignition Facility or the mégajoule laser. This in turn has the potential to enable new nuclear weapons to be developed. The potential effect on global security will be discussed later. In the assessment of risks associated with military research programs one must distinguish the different types of programs currently carried out. I will adopt a classification with regards to the aims pursued.

Three broad classes can be identified:

1. Programs with civilian research objectives i.e. that could be looked after by civilian entities.

   Example: Department of Defence-ID: 04-00861 Next-Generation antibiotics.

   Extracts: "There is a growing national crisis in the treatment of infectious disease driven by the dramatic rise in pathogen multi-drug resistance (MDR)."

   Aim: Identify resistance gene in mobile genetic elements.

2. Programs with "classical" military research objectives i.e. seeking to develop military technologies in continuity with previous ones and that do not rely exclusively on nano-technology.

   Example: Department of Defence – ID: 04-00040, Enhanced Alternative Kinetic Energy Penetrators.

   Aim: "To enhance the penetration capability and behind armour effects of tungsten alloy projectiles via chemistry, mechanical properties and geometry variations, as well as processing innovations or other novel means."

3. "Nano-military" programs contingent on nano effects or properties.

   Example: Central Intelligence Agency – ID: 04-00398 Bio-Nano electronic circuitry

"The combination of biological science with nanotechnology offers the potential to produce a new class of electronic circuitry that rival the complexity and capacity of a human neuro-system"

As expected few risks are associated to the first category of programs, whereas the programs in the third one rise obvious concerns.

The Various Risks

Risks resulting from military nano-research are wide ranging. They include scientific risks, geo-strategic risks, etc. but development of these technologies and their operational deployment also creates military risks. In fact, it is anticipated that the increased complexity of military systems, the speed at which they react, as well as the deployment of thousands of sensors will soon render the battlefield unmanageable by humans.

With the limiting factor becoming the presence in the decision loop of a human operator, the temptation will be strong to remove any human involvement to increase the lethality of the system. This would be a worrying development.

The huge research efforts carried out in the Western world to develop new military applications also creates a dangerous geo-strategic situation. Other countries with significant research capabilities will not stay idle when the wide ranging consensus is that armies relying on nanotechnology will be vastly more effective. As a result one can only expect that the research bias in favour of the defence sector will provoke a new kind of arms race between countries. China and India for example have been increasing their own investments in nano related military research in the past few years.

The Indian prime minister recently declared: "*Indian scientists "must make a break-through in Nanotechnology as it will revolutionise the total concepts of future warfare.*" This in turn will provide justification for continuing or increasing budgets in our countries as disingenuously one could ask: how else are we supposed to defend ourselves from future Chinese nano-weapons?

Nanotechnology also has the potential to threaten existing weapons treaties. Coupled with recent biotech-nology discoveries, nanotech-nology will provide new development options such as targeted bio or chemical weapons, easier delivery methods, etc.

By making chemical and biological weapons easier and "safer" to use, it might provide further incitation to make use of them. On the nuclear weapon side, as noted earlier nanotechnology will most probably help in developing and testing new nuclear weapons thus circumventing the Comprehensive Test Ban treaty (only applicable to bombs >1kT).

Lastly, military funding of research has, in itself, several risks and drawbacks. While this is not specific to nanotechnology and nanosciences the consequences might be quite severe in these fields as they could have a huge potential impact on society.

Military funding of research often puts barriers to the diffusion of scientific results thus producing a lack of transparency. It also prevents proper (independent) risk assessment of research from taking place. Lastly, it tends to steer the advancement of science toward military objectives even when funding non-military fundamental science programs. In fact, as it is a significant source of funding, a researcher facing a choice (due to finite time and resources) of whether to demonstrate the applicability of his field to, say, solar cells or infrared detectors, might be tempted to first investigate the infrared detector as a way to ensure renewal of his DoD grant.

This last point is often overlook but has, I believe, serious deleterious consequences.

Conclusions and Recommendations

The author recommends the immediate transfer of civilian aims programs to existing civil research agencies (NSF, EPSRC, etc.) to put a stop to the presence of a bias in the research process. As a way to ensure transparency of military funding of research the author believes that the list of universities and companies receiving funds from the ministry of defence should be compiled and published on a yearly basis.

A similar list should be published for the new European Defence Agency as it will be funding R&D in addition to its military co-ordination role.

The creation of independent risk assessment bodies (or mechanisms) for nano-military programs is required to ensure that dangerous fields will not be investigated. The author supports Jürgen Altmann's proposed moratorium forbidding the development of non medical intra-body sensors (a technology which could develop massively as a result of nano-technology advancements).

This moratorium could be extended to other areas where, even if the military rational exists to justify development of a technology (it is understandable that a commander wants to know the whereabouts of his troops and the environment they are in), the trickle down effect in civilian spheres would most certainly have a detrimental effect.

In conclusion, nanotechnology (and nanoscience) is a very broad field: it is a new way to see a whole range of subjects. The choice of applications developed will shape the impact this field will have on society (and its perception by the public).

We have to keep in mind that some discoveries require heavy investments but once the knowledge is out it becomes relatively cheap for others to re-discover them. We therefore need to choose responsibly which tech-nologies to develop and which to steer away from.

Military Potential Applications in Nanotechnology

Armies throughout world are looking for a 21st century battle suit.

The clothing must stop bullets, detect chemical, biological agents, monitor a wounded soldier's vital signs, administer basic first aid, and communicate with Headquarters. NT produces stronger, less weight material which is ideal for military applications. Nano-materials (CNT) enable the following material functionalities in uniforms and equipment to make them stronger and lighter could lead to nanofibre-like materials that break off from uniforms and equipment and enter the body and environment.

Light Weight: High strength nano-composite plastics are expected to replace metal and thus reduce weight and Radar signature (RCS) of military weapon platforms. Smart components, with built-in condition and load monitoring sensors improve monitoring of important parameters. Self-repairing or self-healing materials, Shape memory alloys which memorise shape are also possible with NT. Shape memory alloys are used in sleeves of fighter aircraft.

Adaptive Structure: Active structures that adapt to changing conditions such as adaptive camouflage, suspension, flexible/rigid, nanofluids, and shape memory alloys as example of adaptive materials.

Reduce Vibrations: Composites in aircraft also reduce vibrations. In aircrafts particularly in rotator wing suppression of vibration is a great design challenge. NT composite reduces vibrations.

Stealth Nanotechnology

Radar absorption coatings, camouflage and composites reduces RCS. NT is used for developing these materials antiballistic structures, reactive nanoparticle armour, and shock absorbing nanotubes. Nanoparticles are used as surface coverings to make it harder, smoother, and/or stealthier. IR image also reduces with composits.

Filter: Nanomaterials used in filters to remove selected impurities from fluids could become very low in cost and hence ubiquitous, and result in many small but discrete concentrations of possibly toxic impurities. "Argonide Nanomaterials" filters can be used in war where water contamination is biggest threat.

Applications Under Development

Body Armor made of CNT is 117 times stronger than steel, these can be used for bullet proof, helmets and safety device bullet-proof vests are basically made from high stiffness and toughness, woven or laminated, polymeric fibres stacked in a number of layers. Upon impact of the striking bullet, the fabric material absorbs the energy by stretching of the fibres and the stiff fibres ensure that the load is dispersed over a large area throughout the material. CNT is an ideal candidate material for bulletproof vests due to its unique combination of exceptionally high elastic modulus and high yield strain. If one compares these values with those for other fibres suitable for ballistic applications, the enormous potential of CNTs as a candidate material for bullet-proof armor system is quite evident. *Abalone shell* can be used as body armour.

Armour Vehicle

Armour Vehicles uses Al2O3, which is standard high performance form a grain size of 2-10 micro-metre; Vickers hardness of 1,750 (HV 10); a binding strength of 385 MPa and elastic modulus of 410 GPa. Nano crystalline version of Al2O3 , Silicon carbide (SiC), with grain size of 50 nm and 20 nm respectively. This offer hardeners value 3,000 with 70/840 MPa and elastic moduli of 680/900 GPa. "Nanodyne" makes a tungsten-carbide-cobalt composite powder (grain size less than 15nm) that is used to make a sintered alloy as hard as diamond, which is in turn used to make cutting tools, drill bits, armour plate, and jet engine parts.

Nano Fibre

Primitive NBC suit reduces solider capability because it can't pass air. NT in cloths nonwoven composite fabrics can be made 25% lighter and stronger than current woven military uniform fabrics. In addition, they can be made to exhibit over three times the air permeability or breathability of the current woven military uniform fabrics. Uniform fabric will offer enhanced abrasion resistance, develop a fire and chemical/biological resistant fabric and offer combination water repellent/moisture absorbent composite fabric. Nano-silver socks/undergarments will remain hygienic for days in war scenario, where it is impossible to change undergarments.

Nano Food

NT enable food will last more without getting rotten, these foods are ideal for war where supply of food is difficult. Nano-Silver packed food withstand days together without damage. These foods can be used in disaster recovery teams, in war and hash areas like very high altitude areas where fresh food is a difficulty.

Nano Energetic Materials (nEMs)

Nanoscale energetic materials (*nEMs*) offers much higher energy densities, faster rate of energy release, greater stability, and more security. Batteries used for combat gadgets with *nEMs* definitely have much less weight and a relief to soldiers. Nano-inspired batteries will provide reliable power to military weaponry while requiring less volume themselves. Shrinking batteries will enable devices to shrink in size. nEMs reduce battery weight by 20% to 30% which mean soldiers to carry less weight in war.

Future Applications

The military use of NT should lead to higher protection, more lethality, longer endurance and better self-supporting capacities of future soldiers. Water purifier kit, which will be the part of survival kit in jungle operation to clean water to drink.

Electro Chromic Camouflage

With NT electro chromic camouflage can be achieved, by which soldiers can disappear to, seen by necked eye. Fabric made of Electro-chromic camouflage, which changes colours instantly to blend in with the surroundings. Nano enabled paint currently developing a special paint that makes drones, missiles, or aircraft simply disappear, or to be more precise, they become very difficult to detect. NT developed envelopes the object, absorbs the radio waves emitted by the radar, and releases them as heat energy scattered in space. In doing so, the material disguises the object, making it difficult to identify by Radar.

Fabrics

Dust repellent /strain resistance/ fire resistance cloths for warfare. Nanopolymers developed for soldier act as fabric to breakdown bio/chemical warfare agents. Biosensors can be used to monitor a soldier's health. Nano sized silicon carbide particles for physical protection.

Condition Monitoring

Modern maintenance philosophy is based on on-line monitoring and e-maintenance. Nanosensors are engraved in structure of aircraft. Important flight parameters can be monitored at base station even in case of aircraft is flying from one place to another.

War Tag

Smart NT alloy with RFID is proposed as war tag with nanosensor, presently, Indian Armed Forces soldiers uses metallic piece on which name, number and religion is engraved. In miniaturization of gadgets like RFID/ search and rescue mission, nano sensors embedded in combat suit can emit signal which can be picked by search and rescue operation.

Artificial Muscles

Enhancing soldier capabilities with artificial muscles that could enable soldiers to leap tall walls, if wounded or required, uses nanotechnology and electricity. It flexes when jolted by electricity, and then relaxes when the electricity is turned off. So far, though, its reactions are much too slow.

- Nano enabled medicine for solders, for and curing crack bones in field without plaster. Field ambulance units of Armed Forces should carry these instant NT enable medicine for war. If a soldier is injured in the arm or leg, nano-fibres in the uniform, the fabric would constrict into a tourniquet. This will be a real life-saver, because half of all battlefield deaths are due to massive blood loss before wounded soldiers can be treated.
- Composite materials made of NT having no metals, light weight and more strength with low RCS in military aircraft can make military aircraft more lethal. Aircraft with NT materials have more thrust to weight ratio, less maintenance problems and more lethal. MEMS (Micro Electro Mechanical Sensors) reduce the overall size of equipment, now with NT it is NEMS (Nano Electro Mechanical Sensors) and OLED (Organic Light Emitting Diode); the future of television, as electricity passes over the thin nono-polymer film layer it produces colour and light. More Dots of lights mean a better picture, less power consumption. Displays will be better and small with less power consumption. Modern armies with electronic gadgets want less weight to carry by solders.

NT will help to reduce the weight of electronics but also helps in, fast processing, better display and sleek dimension of palm tops, laptops or PDA by NT.

Research Area of Nano Sensor

Nano Air Vehicle (NAV) aims at an extremely small (less than 7.5 cm wing span), ultra-lightweight (less than 10 grams) air vehicle system, designed for indoor and outdoor urban military missions. "Smart dust" is a device which contains tiny wireless Micro Electro Mechanical Sensors (MEMS) that can detect everything from light to vibrations. "Mote" is a autonomous sensor with power source, processor, Tiny Operating system and Antenna. They form Adhoc networks in tactical / disaster area. "Mote" weighs not more than 10 grams and can carry a payload of up to 2 grams. The will help protect the lives and enhance the operational effectiveness of soldiers and first responders in case of disaster.

Nano Poisons

Most people think of poison as a tool for killing tool, but NT, with its ability to trigger specific brain functions, will provide a whole new menu of poison options. As an example, a liar-poison will make it impossible for someone to tell the truth. A *kleptomaniac* poison will make it impossible for the person to stop stealing things. An alcoholic poison will make a person unable to stop drinking alcohol. The obesity poison will cause a person to eat themselves to death, and favourite the "frontal lobotomy poison," that will make a person incapable of being angry or mean.

NT for In-Capacitative Agents

Neutralizing a person's memory can often be a more powerful defence than killing them. Micro fields flaring up in a succession of unnoticeable tiny brain bursts may wipe sections of a brain clean without anyone ever noticing.

Nano Needles

Invisible to the human eye, nano diameter needles will be shot like clusters of bullets from great distances to "pin" people to a wall or freeze their physical movement. Nano needles, because of their tiny diameter, will be the ultimate non-lethal weapon, leaving no visible wounds and causing no permanent damage. This may be used as arrest intruder without harming or killing him.

Water Bullets

As a different kind of non-lethal weapon, self-contained water balls, formed around an elevated surface tension containment system, will be used to knock people down, and temporarily rendering them harmless.

Desynchronized Energy Fields

Binary power, created by the intersection of two otherwise harmless beams, has the ability to disrupt the energy fields in an individual. A person with desynchronized energy fields will feel extremely fatigued, and pushed to a more extreme level, will drop unconsciously to the ground. New form of stun-gun with NT.

Proposed Application

NT enable survival kit is a part of ejection seat in fighter planes. For survival of pilot, military aircraft contains survival kit which is used in case of ejection. The survival kit contains essential items like (knife, water purifier tablets, high calories chock lets and pistol). It is proposed to modernise the contents of survival kit with following:

- Nano enabled high calories eatables, Nano bandage, Nano medicine (nano-bio fusion), Nano water purifier, nanosensor (RFID, tag).
- Weapon and power source with NT reduces weight for solider to carry by 20-30%. PDA and watch which will communicate with Net centric warfare control room.
- NT enable war tag with RFID, so that it can be used by rescue and search also.

Aircraft Canopy

Aircraft canopy can be coated with CNT which will not only give strength but also reduces Electromagnetic Interference (EMI) from space which is dangerous for pilot as well as for avionics electronics. "Argonide" using aluminum nanoparticles, argonide has created rocket propellants that burn at double the rate. They also produce copper nanoparticles that are incorporated into automotive lubricant to reduce engine wear.

NT in Aviation

CNT polymer composites have high Young's modulus, high thermal performance and crash resistance, some of CNT for aircraft structure are CNT/polyimide, CNT epoxy, and CNT/PP. Aero engine performance can be enhanced by SiC nanoparticles, SiC particle reinforced alumina. TiN nanocrystallites embedded in amorphous Si3N4 are used for Wear-resistant coatings. Nanographide nanosilica is used in gaskets for aircraft. Nano fabrics are used for Arrester barrier and tail parachute for fighter aircraft. Nanochromium is used for corrosion resistant.

UCAV (Un-Manned Combat Aerial Vehicle)

NT enabled UCAV are very small with low RCS as most of the portions are made of composites. NT enable weapons are too small and more lethal as compared to conventional weapon. These pilotless UCAV can be deployed in Hostile terrain and adverse weather when fighter flying in not passable.

NT is multidisciplinary; it is used for improving existing technologies and exploring new compounds. Using NT to create materials with properties that will revolutionize military technology, from processors to display screens and body armor to air filters. This is a beginning and still lot to do, however there are potential applications in military technology.

A future of nano-dust spies, sentinels, and defencive weapons, is one that most of the world's military specialists are unprepared. Yet most of these weapons are far closer and easier to devise and build, than the molecular nano-assembler, the horn of plenty that most people think of when they think of a future NT.

Nano-weapons combined with bio/chemical and genetic weapons; provide the budding world religious or ideological dictator with far more ultimate power than a few nuclear weapons.

Every measure has a counter-measure. But not everyone will have the resources to obtain counter-measures, when the means of deadly attack becomes nearly ubiquitous.

Soldier capabilities can be enhanced by NT with strong, lightweight materials for soldier systems and system components adaptive, multifunctional materials for soldier systems and system components novel detection and protection schemes for bio/chemical warfare threats and identification of friend or foe. Remote and local soldier monitoring systems, wound and injury triage and emergency treatment systems to enhance soldier survivability novel non-combat and combat performance enhancement systems for the soldier system that would improve soldier survivability.

Still very little is known about NT and its side effects. It is unknown if producing items this small could be dangerous to the manufacturers or the soldiers using them for protection, as it is not known if the nanoscale devices could penetrate cell membranes and skin on the human body. As society continues, research will continue to be done, and hopefully the possibilities can do more protecting and helping than damaging as we enter in new "nano age".

It is proposed to modernise old systems with NT enable systems. Design of military weapons and platform should incorporate NT in each aspect for miniaturization effectiveness, less weight and more effectiveness. Old systems can replace subsystem with NT enable sub-system for enhance operation capability like survival kit modernisation, display system of aircraft. NT enable combat suit, systems and weapons are need of the hour for Armed Forces.

Milestones in Nanotechnology

The nano-level world is an entirely different place where materials with familiar properties take on new and unfamiliar properties. At this level of scale, often referred to as the mesoscale, physical properties are subject to a combination of rules of quantum and classical physics.

Basic Definitions and Concepts

Nanotechnology is still an emerging technology and the nomenclature and classification of the materials is yet to be developed. The American National Standards Institute (ANSI) is in the process of establishing a standardized nomenclature for nanomaterials and has formed a Nanotechnology Standards Panel which first assembled in late September 2004. For now, the basic concepts in nanotechnology may be known by multiple names.

Buckyball

A buckyball, or fullerene, is named after Buckminster Fuller, a futurist and global thinker best known for the invention of the geodesic dome. The structure consists of 60 carbon atoms arranged in a sphere with a pattern of hexagons and pentagons similar to a soccer ball. The buckyball is hollow, but can be used to carry another material; the outside surface of the buckyball can also be coated. A buckyball is also a semiconductor and small enough to penetrate biomembranes ordinarily thought relatively impenetrable such as cell walls or the blood brain barrier. Given these unique properties, it is not surprising that buckyballs are source of great interest to engineers and pharmaceutical researchers. The discovery of the fullerene was important enough to earn Richard Smalley, Robert Curl and Sir Harold Kroto the 1985 Nobel Prize for chemistry.

Functioning of STM

An STM brings a sharp, electrically conducting needle up to an electrically-conducting surface, almost touching it. The needle and surface are electrically connected, so that a current will flow if they touch, like closing a switch. But at just what point do soft, fuzzy atoms "touch"?

It turns out that a detectable current flows when just two atoms are in tenuous contact-fuzzy fringes barely overlapping-one on the surface and one on the tip of the needle. By delicately maneuvering the needle around over the surface, keeping the current flowing at a tiny, constant rate, the STM can map shape of the surface with great precision. Indeed, to keep the current constant, the needle has to go up and down as it passes over individual atoms.

The STM was invented by Gerd Binnig and Heinrich Rohrer, research physicists studying surface phenomena at IBM's research labs in Zurich, Switzerland. After working through the 1970s, Rohrer and Binnig submitted their first patent disclosure on an STM in mid-1979. In 1982, they produced images of a silicon surface, showing individual atoms. Ironically, the importance of their work was not immediately recognised: Rohrer and Binnig's first scientific paper on the new tool was rejected for publication on the grounds that it was "not interesting enough." Today, STM conferences draw interested researchers by the hundreds from around the world.

In 1986-quite promptly as these things go-Binnig and Rohrer were awarded a Nobel Prize. The Swedish Academy explained its reasoning: "The scanning tunneling microscope is completely new and we have so far seen only the beginning of its development. It is, however, clear that entirely new fields are opening up for the study of matter." STMs are no longer exotic: Digital Instruments of Santa Barbara, California, sells its system (the Nanoscope) by mail with an atomic-resolution-or-your-money-back guarantee. Within three years of their commercial introduction, hundreds of STMs had been purchased.

Working of an AFM

The related atomic force microscope is even simpler in concept: A sharp probe is dragged over the surface, pressed down gently by a straight spring. The instrument senses motions in the spring (usually optically), and the spring moves up and down whenever the tip is dragged over an atom on the surface. The tip "feels" the surface just like a fingertip in the simulated molecular world. The AFM was invented by Binnig, Quate, and Gerber at Stanford University and IBM San Jose in 1985. After the success of the STM, the importance of the AFM was immediately recognised. Among other advantages, it works with nonconducting materials. The next chapter will describe how AFM-based devices might be used as molecular manipulators in developing molecular nanotechnology. As this is written, AFMs have just become commercially available.

Can Proximal Probes Move Atoms?

To those thinking in terms of nanotechnology, STMs immediately looked promising not only for seeing atoms and molecules but for manipulating them. This idea soon became widespread among physicists. As Calvin Quate stated in Physics Today in 1986, "Some of us believe that the scanning tunneling microscope will evolve... that one day [it] will be used to write and read patterns of molecular size." This approach was suggested as a path to molecular nanotechnology in Engines of Creation, again in 1986. By now, whole stacks of scientific papers document the use of STM and AFM tips to scratch, melt, erode, indent, and otherwise modify surfaces on a nanometre scale. These operations move atoms around, but with little control. They amount to bulk operations on a tiny scale-one fine scratch a few dozen atoms wide, instead of the billions that result from conventional polishing operations.

In 1987, R. S. Becker, J. A. Golovchenko, and B. S. Swartzentruber at AT&T Bell Laboratories announced that they had used an STM to deposit small blobs on a germanium surface. Each blob was thought to consist of one or a few germanium atoms. Shortly thereafter, IBM Almaden researchers John Foster, Jane Frommer, and Patrick Arnett achieved a milestone in STM-based molecular manipulation. Of this team, Foster and Arnett attended the First Foresight Conference on Nanotechnology, where they told us the motivations behind their work.

Foster came to IBM from Stanford University, where he had completed a doctorate in physics and taught at graduate school. The STM work was one of his first projects in the corporate world. He describes his colleague Arnett as a former "semiconductor jock" involved in chip creation at IBM's Burlington and Yorktown locations. Besides his doctorate in physics, Arnett brought mechanical-engineering training to the effort.

Arnett explains what they were trying to do: "We wanted to see if you could do something on an atomic scale, to create a mechanism for storing information and getting it back reliably." The answer was yes. In January 1988, the journal Nature carried their letter reporting success in pinning an organic molecule to a particular location on a surface, using an STM to form a chemical bond by applying an electrical pulse through the tip. They found that having created and sensed the feature, they could go back and use another voltage pulse from the tip to change the feature again: enlarging it, partly erasing it, or completely removing it. IBM quickly saw a commercial use, as explained by Paul M. Horn, acting director of physical sciences at the Thomas J. Watson Research Centre: "This means you can create a storage element the size of an atom. Ultimately, the ability to do that could lead to storage that is ten million times more dense than anything we have today." A broader vision was given by another researcher, J. B. Pethica, in the issue of Nature in which the work appeared: "The partial erasure reported by Foster et al. implies that molecules may have pieces deliberately removed, and in principle be atomically 'edited,' thereby demonstrating one of the ideals of nanotechnology."

Foster's group succeeded in pinning single molecules to a surface, but they couldn't control the results-the position and orientation-precisely. In April 1990, however, another group at the same laboratory carried the manipulation of atoms even further, bringing a splash of publicity. Admittedly, the story must have been hard to resist: it was accompanied by an STM picture of the name IBM," spelled out with thirty-five precisely placed atoms. The precision here is complete, like the precision of molecular assembly: each atom sits in a dimple on the surface of a nickel crystal; it can rest either in one dimple or in another, but never somewhere between.

Donald Eigler, the lead author on the Nature paper describing this work sees clearly where all this is leading: "For decades, the electronics industry has been facing the challenge of how to build smaller and smaller structures. For those of us who will now be using individual atoms as building blocks, the challenge will be how to build up structures atom by atom."

How Far can Proximal Probes take Us?

Proximal probes have advantages as a tool for developing nanotechnology, but also weaknesses. Today, their working tips are rough and irregular. To make stable bonds form, John Foster's group used a pulse of electricity, but the results proved hard to control. The "IBM" spelled out by Donald Eigler's group was precise, but stable only at temperatures near absolute zero-such patterns vanish at room temperature because they are not based on stable chemical bonds. Building structures that are both stable and precise is still a challenge. To form stable bonds in precise patterns is the next big challenge.

John Foster says, "We're exploring a concept which we call 'molecular herding,' using the STM to 'herd' molecules the way my Shetland sheep dog would herd sheep... Our ultimate goal with molecular herding is to make one particular molecule move to another particular one, and then essentially force them together. If you could put two molecules that might be small parts of a nanomachine on the surface, then this kind of herding would allow you to haul one of them up to the other. Instead of requiring random motion of a liquid and specific chemical lock-and-key interactions to give you exactly what you want in bringing two molecules together [as in chemical and biochemical approaches], you could drive that reaction on a local level with the STM. You could use the STM to put things where you want them to be." The next chapter will discuss additional ideas for using proximal probes in early nanotechnology.

Proximal-probe instruments may be a big help in building the first generation of nanomachines, but they have a basic limit: Each instrument is huge on a molecular scale, and each could bond only one molecular piece at a time. To make anything large-say, large enough to see with the naked eye-would take an absurdly long time. A device of this sort could add one piece per second, but even a pinhead contains more atoms than the number of seconds since the formation of Earth. Building a Pocket Library this way would be a long-term project.

How can Such Slow Systems Ever Build Anything Big?

Rabbits and dandelions contain structures put together one molecular piece at a time, yet they grow and reproduce quickly. How? They build in parallel, with many billions of molecular machines working at once. To gain the benefits of such enormous parallelism, researchers can either (1) use proximal probes to build a better, next-generation technology, or (2) use a different approach from the start. The techniques of chemistry and biomolecular engineering already have enormous parallelism, and already build precise molecular structures. Their methods, however, are less direct than the still hypothetical proximal probe-based molecule-positioners.

They use molecular building blocks shaped to fit together spontaneously, in a process of self-assembly. David Biegelsen, a physicist who works with STMs at the Xerox Palo Alto Research Centre, put it this way at the nanotechnology conference: "Clearly, assembly using STMs and other variants will have to be tried. But biological systems are an existence proof that assembly and self-assembly can be done. I don't see why one should try to deviate from something that already exists.

Chemical Mixing and Building Molecular Objects

An engineer would say that chemists (at least those specialising in synthesis) are doing construction work, and would be amazed that they can accomplish anything without being able to grab parts and put them in place. Chemists, in effect, work with their hands tied behind their backs.

Molecular manufacturing can be termed "positional chemistry" or "positional synthesis," and will give chemists the ability to put molecules where they want them in three-dimensional space. Rather than trying to design puzzle pieces that will stick together properly by themselves when shaken together in a box, chemists will then be able to treat molecules more like bricks to be stacked. The basic principles of chemistry will be the same, but strategies for construction will become far simpler.

Without positional control, chemists face a problem something like this: Picture a giant glass barrel full of tiny battery-powered drills, buzzing away in all directions, vibrating around in the barrel. Your goal is to take a piece of wood and put a hole in just one specific spot. If you simply throw it in the barrel, it will be drilled haphazardly in many places. To control the process, you must protect all the places you don't want drilled-perhaps by gluing protective pieces of metal over most of the wood surface. This problem-how to protect one part of a molecule while altering another part-has forced chemists to develop ever-cleverer ploys to build larger and larger molecules.

Chemists can achieve great things, but have focused much of their effort on duplicating molecules found in nature and then making minor variants. As an example, take palytoxin, a molecule found in a Hawaiian coral. It was so difficult to make in the lab that it has been called "the Mount Everest of synthetic chemistry," and its synthesis was hailed as a triumph. Other efforts are poured into making small molecules with unusual bonding, or molecules of remarkable symmetry, like "cubane" and "dodecahedrane" (shaped like the Platonic solids they are named after).

Chemists, at least in the United States, regard themselves as natural scientists even when their life's work is the construction of molecules by artificial means. Ordinarily, people who build things are called engineers. And indeed, at the University of Tokyo the Department of Synthetic Chemistry is part of the Faculty of Engineering; its chemists are designing molecular switches for storing computer data. Engineering achievements will require work directed at engineering goals.

The Move towards Building Molecular Machines

Molecular engineers working toward nanotech-nology need a set of molecular building blocks for making large, complex structures. Systematic building-block construction was pioneered by Bruce Merrifield, winner of the 1984 Nobel Prize in Chemistry. His approach, known as "solid phase synthesis," or simply "the Merrifield method," is used to synthesize the long chains of amino acids that form proteins.

In the Merrifield method, cycles of chemical reaction each add one molecular building block to the end of a chain anchored to a solid support. This happens in parallel to each of trillions of identical chains, building up trillions of molecular objects with a particular sequence of building blocks. Chemists routinely use the Merrifield method to make molecules larger than palytoxin, and related techniques are used for making DNA in so-called gene machines: an ad from an Alabama company reads, "Custom DNA-Purified and Delivered in 48 hours."

While it's hard to predict how a natural protein chain will fold-they weren't designed to fold predictably-chemists could make building blocks that are larger, more diverse, and more inclined to fold up in a single, obvious, stable pattern. With a set of building blocks like these, and the Merrifield method to string them together, molecular engineers could design and build molecular machines with greater ease.

4

Applications of Nanotechnology in Military, Space and Security

Background

In the USA, which can boast the most government financial support for nanotechnology research in the world, nanotechnology funding given by the federal government to the Department of Defence is second only to that given to the National Nanotechnology Initiative (NNI). The Massachusetts Institute of Technology (MIT) hosts the US Army Institute of Soldier Nanotechnologies (ISN), a research unit devoted to developing military applications for nanotechnology, its aim being to improve the "survival of the soldier of the future".

Using Smart Materials to Improve Body Armour

The Institute's ultimate goal is to "create a 21st century battlesuit". To this end they are investigating smart materials able to be responsive to the conditions of their environment, sensors able to detect chemical or biological warfare agents, and lightweight bullet-proof materials. There are also attempts to incorporate wound detection and treatment systems within uniforms. For example, responsive systems, such as the material hardening to provide an instant splint for a broken bone, are in development. The five-year contract awarded in 2002 to MIT by the US Army implies the time-scale for the development of these projects is rather near term. Lighter and stronger weapons and equipment are being developed, although information on this area is scarce and it is assumed that much of this research is classified.

Using Nanotechnology to Improve the Human-Machine Interface

In addition there is a focus on improving the human-machine interface, perhaps including sensory enhancement such as direct retinal displays or communication to the ear. Should these products be developed, and be able to be manufactured, it is likely that they will be transferred to the space programme and eventually into wider usage.

Using Improved Materials in the Building of Spaceships

Improved materials, lighter but with tough, heat resistant properties, are being used in the design and construction of spacecraft and satellites, and this process will gain from nanotechnology. (Confusingly, the term nanosatellite is in use but refers to a small satellite, not one using nanotechnology.)

Improving Security Using Nano-Based Sensors, Nanoparticles and DNA Nanotechnology

There is also the possibility of nanotechnology facilitating improvements in civilian security equipment. The Institute of Nanotechnology suggests fingerprinting will become cheaper, quicker and more effective using DNA techniques involving nanotechnology, and there is also the possibility that nano-based sensors could be used as electronic detectors ('sniffer dogs') for improved airport security. Quantum dots, fluorescent nanoparticles which glow when exposed to ultraviolet light, may be used as tags and labels to prevent theft and counterfeiting, and to trace the course of drugs within the body.

Military Uses of Nanotechnology

Another major political issue to be considered in the balance of power is how nanotechnologies may affect the military and the future arms race. Augmenting human performance has obvious attractions, such as improving stamina, strength and the ability to deal with multiple situations and process information more effectively.

Technological Advances and the Military

History also shows us that technology advances have lead to the military developing smaller, more mobile and more powerful weapons, surveillance systems and other military equipment. While these may be developed under the auspices of safeguarding freedom and ensuring security, such devices if they continue to follow current trends could give rise to an arms race similar to that of nuclear weapons.

Regulations to Prevent an Arms Race

To prevent an escalation there must be internationally agreed regulations. For instance, nanotechnology could one day make surveillance devices and weapons that are virtually undetectable and have advanced remote control capabilities, which raises issues of where the dividing line lies between ensuring security and infringing human rights.

As a result, unilateral actions of one country on another could become more targeted and more difficult for the international community to police. Safe-guarding these technologies will be crucial-preventing their sale or distribution to, or theft by, other organisations.

Military-Based Nanotechnology Programmes in the United States

At present the main country publicly declaring that it is developing military uses of nanotechnologies is the US. In 2001 the US invested 125 million USD in nanotechnology initiatives for the Department of Defence (DoD). The proposed budget for 2005 includes 276 million USD for the DoD and 1 million USD for homeland security (together this is approximately 28% of the total nanotechnology budget).

For example, the Institute of Soldier Nanotechnologies (ISN) at MIT was established in 2002 with a five-year 50 million USD contract from the US military and is looking at producing a battle-suit that provides improved protection, monitors vital signs and (at the same time) is lighter than conventional equipment.

Military-Based Nanotechnology Programmes in the United Kingdom

The UK Ministry of Defence (MOD) is also funding some military aspects of nanotechnology including new structural materials, electronic devices and quantum interference, however this is on a much smaller scale (approximately £1.5 million per annum). This research is mainly carried out through the Defence Evaluation and Research Agency (DERA) and its Corporate Research Programme, although the MOD recognises that close collaboration with academic research is essential to keep abreast of developments in nanotechnology that could have applications in defence.

Military-Based Nanotechnology Programmes in Sweden

Sweden is investing 11 million euros over 5 years in nanotechnology research for military purposes through FOI (the Swedish Defence Research Agency). The funding is divided between 7 research projects (involving universities, major defence companies and some high-tech start-up companies) and focussed during the first two years on feasibility studies before entering an application-oriented phase.

Military-Based Nanotechnology Programmes in the European Union

The EU has recently published a call for proposals and supporting activities in the scope of the preparatory action on "the enhancement of the European industrial potential in the field of security research". While this does not specifically mention nanotechnologies, many of the areas have implicit nanotechnology connections (such as sensors, tracking devices) under the project themes of "Improving situation awareness" and "Protecting against terrorism (including bio-terrorism and incidents with biological, chemical and other substances)". There is a budget of 65 million euros from 2004-2006, which is a precursor to a much larger Security Research Programme that will start in 2007.

Nanotechnology holds promise to create more advanced and refined weaponry in the military, according to a report about the work of scientists in the U.S. Department of Energy's Ames Laboratory. Armor-piercing projectiles, usually manufactured using uranium, could potentially be developed using a Nanocomposite material made of tungsten and glass. The need for replacing Uranium arises out of environmental concerns. Uranium emits radiation as it depletes, and can also create a toxic environment for soldiers who must stockpile the Uranium-based munitions.

A projectile needs to be more than environmentally friendly in order to function as an effective weapon, however. In particular, it needs to shear easily. Shearing means that upon impact, the projectile will gradually "sharpen itself" as layers of metal are stripped away and it becomes as pointed as a finely pointed spear. This is necessary for the weapon to forcefully penetrate enemy armor. These requirements have made high density Uranium ideal for ballistic applications.

However, scientists have found an effective substitute in the form of tungsten and metallic glass that can be used to build a projectile. Tungsten is slightly more dense than Uranium, and glass has been shown to shear more easily. However, composites made of tungsten and glass create a "mushroom head" effect upon impact, eliminating the propensity of the projectile to shear and thus nullifying its effectiveness. This is where Nanotechnology comes in.

Scientists have discovered that when the tungsten and glass are reduced to the Nanometre scale, they shear much more easily. Researchers have been able to create Nanoscale composites without having either the glass or the tungsten blend as alloys. Each element retains its property and the projectile shears effectively. Work continues in military applications of Nanotechnology and no doubt more discoveries will be unveiled in the near future.

Preventing Errors in Molecular Manufacturing

Many people familiar with today's manufacturing technologies will assume that such a thing is near-impossible. Today's manufacturing operations are doing very well to get one error in a million products. To reduce the error to one in a billion—to say nothing of one in a million billion, which Drexler talks about in *Nanosystems*—seems ridiculously difficult. None of today's technologies could do it, despite many decades of improvement. So how can molecular manufacturing theorists reasonably expect to develop systems with such low error rates—much less, to develop them on a schedule measured in years rather than decades?

There are physical systems today that have error rates even lower than molecular manufacturing systems would require. A desktop computer executes more than a billion instructions each second, and can operate for years without a single error. Each instruction involves tens of millions of transistors, flipping on or off with near-perfect reliability. If each transistor operation were an atom, the computer would process about a gram of atoms each day—and they would all be flawlessly handled.

The existence of computers demonstrates that an engineered, real-world system, containing millions of interacting components, can handle simple operations in vast numbers with an insignificant error rate. The computer must continue working flawlessly despite changes in temperature, line voltage, and electromagnetic noise, and regardless of what program it is asked to run.

The computer can do this because it is digital. Digital values are discrete—each signal in the computer is either on or off, never in-between. The signals are generated by transistor circuits which have a non-linear response to input signals; an input that is anywhere near the ideal "on" or "off" level will produce an output that is closer to the ideal. Deviations from perfection, rather than building up, are reduced as the signal propagates from one transistor to another. Even without active error detection, correction, or feedback, the non-linear behaviour of transistors means that error rates can be kept as low as desired: for the purposes of computer designers, the error rate of any signal is effectively zero.

An error rate of zero means that the signals inside a computer are perfectly characterized: each signal and computation is exactly predictable. This allows a very powerful design technique to be used, called "levels of abstraction." Error-free operations can be combined in intricate patterns and in large numbers, with perfectly predictable results. The result of any sequence of operations can be calculated with certainty and precision.

Thousands of transistors can be combined into number-processing circuits that do arithmetic and other calculations. Thousands of those circuits can be combined in general-purpose processor chips that execute simple instructions. Thousands of those instructions can be combined into data-processing functions. And those functions can be executed thousands or even billions of times, in any desired sequence, to perform any calculation that humans can invent... performing billions of billions of operations with reliably zero errors.

Modern manufacturing operations, for all their precision, are not digital. There is no discrete line between a good and a bad part—just as it's impossible to say exactly when someone who loses one hair at a time becomes bald. Worse, there is no mechanism in manufacturing that naturally restores precision. Difficult and complicated processes are required to construct a machine more precise than the machines used to build it. To build a modern machine such as a computer-controlled lathe requires so many different techniques—polymer chemistry, semiconductor lithography, metallurgy and metal working, and thousands of others—that the "real world" will inevitably create errors that must be detected and corrected. And to top it off, machines suffer from wear—their dimensions change as they are used.

Given the problems inherent in today's manufacturing methods and machine designs, the idea of building a fully automated general-purpose manufacturing system that could build a complete duplicate of itself... is ridiculous.

The ability to form covalent solids by placing individual molecules changes all that. Fundamentally, covalent bonds are digital: two atoms are either bonded, or they are held some distance apart by a repelling force. (Not all bond types are fully covalent, but many useful bonds including carbon-carbon bonds are.) If a covalent bond is stretched out of shape, it will return to its ideal configuration all by itself, without any need for external error detection, correction, or feedback.

If a covalent-bonding manufacturing system performs an operation with less than one atom out of place, then the resulting product will have exactly zero atoms out of place. Just like transistor signals in a digital computer, imperfections fade away all by themselves. (In both cases, a bit of energy is used up in making the imperfections disappear.) In digital systems, there is no physical law that requires imperfections to accumulate into errors—not in digital computer logic, and not in atomic fabrication systems.

Atomic fabrication operations, like transistor operations, can be characterized with great reliability. Only a few transistor operations are a sufficient toolkit with which to design a computer. A general-purpose molecular manufacturing system may use a dozen or so different kinds of atoms, and perhaps 100 reactions between the atoms. That is a small enough number to study each reaction in detail, and know how it works with as much precision as necessary. Each reaction can proceed in a predictable way each and every time it is attempted.

A sequence of completely predictable operations will itself have a completely predictable outcome, regardless of the length of the sequence. If each one of a sequence of a billion reactions is carried out as expected, then the final product can be produced reliably.

Chemists who read this may be objecting that there's no such thing as a reaction with 100% yield. Answering that objection in detail would require a separate essay—but briefly, mechanical manipulation and control of reactants can in many cases prevent unwanted reaction pathways as well as shifting the energetics so far (hundreds of zJ/bond or kJ/mole) that the missed reaction rate is reduced by many orders of magnitude.

At this point, it is necessary to consider the "real world." What factors, in practice, will reduce the predictability of mechanically-guided molecular reaction machinery?

One factor that doesn't have to be considered in a well-designed system of this type is wear. Again, it would take a separate essay to discuss wear in detail, but wear in a covalent solid requires breaking strong inter-atomic bonds, and a well-designed system will never, in normal operation, exert enough force on any single atom to cause its bonds to break. Likewise, machines built with the same sequence of reliable operations will themselves be identical. Once a machine is characterized, all of its siblings will be just as fully understood.

Mechanical vibration from outside the system is unlikely to be a problem. It is a problem in today's nanotech tools because the tools are far bigger than the manipulations or measurements those tools perform—big enough to have slow vibrational periods and high momentum. Nanoscale tools, such as would be used in a molecular manufacturing system, would have vibrational frequencies in the gigahertz or higher, and momentum vanishingly small compared to restoring forces.

It is possible that vibrations generated within the system, from previous operations of the system or of neighbouring systems, could be a problem. In computers, transistor operations can cause voltage ripples that cause headaches for designers, and are probably analogous. But these problems are practical, not fundamental.

Contaminant molecules should not be a problem in a well-designed system. The ability to build covalent solids without error implies the ability to build hermetically sealed enclosures. Feedstock molecules would have to be taken in through the enclosures, but sorting mechanisms have been planned that should reject any contaminants in the feedstock stream with extremely low error rates. There are ways for a manufacturing system inside a sealed enclosure to build another system of the same size or larger without breaking the seal. It would take a third essay to discuss these topics in detail, but they have been considered and none of the problems appears unlikely to be addressable in practice.

Despite everything written above, there will be some fraction of molecular manufacturing systems that suffer from errors—if nothing else, background levels of ionizing radiation will cause at least some bond breakage. In theory an imperfect machine could fabricate more imperfect machines, perpetuating and perhaps exacerbating the error. However, in practice, this seems unlikely Whereas a perfect manufacturing machine could do a billion operations without error, an imperfect machine would probably make at least one atom-placement error fairly early in the fabrication sequence. That first error would leave an atom out of its expected position on the surface of the workpiece. A flawed workpiece surface would usually cause a cascade of further fabrication errors in the same product, and long before a product could be completed, the process would be hopelessly jammed. Thus, imperfect machines would quickly become inert, before producing even one imperfect product.

The biggest source of unpredictability probably will be thermal noise, sometimes referred to as Brownian motion. (Quantum uncertainty and Heisenberg uncertainty are similar but smaller sources of unpredictability.) Thermal noise means that the exact dimensions of a mechanical system will change unpredictably, too rapidly to permit active compensation. In other words, the exact position of the operating machinery cannot be known. The degree of variance depends on the temperature of the system, as well as the stiffness of the mechanical design. If the position varies too far, then a molecule-bonding operation may result in one of several unpredictable outcomes.

This is a practical problem, not a fundamental limitation; in any given system, the variance is limited, and there are a number of ways to reduce it. More research on this point is needed, but so far, high-resolution computational chemistry experiments by Robert Freitas seem to show that even without using some of the available tricks, difficult molecule-placement operations can be carried out with high reliability at liquid nitrogen temperatures and possibly at room temperature. If positional variance can be reduced to the point where the molecule is placed in approximately the right position, the digital nature of covalent bonding will do the rest.

This is a key point: *The mechanical unpredictability in the system does not have to be reduced to zero, or even extremely close to zero, in order to achieve extremely high levels of reliability in the product.* As long as each reaction trajectory leads closer to the right outcome than to competing outcomes, the atoms will naturally be pulled into their proper configuration each time—and by the time the next atoms are deposited, any positional error will have dissipated into heat, leaving the physical bond structure perfectly predictable for the next operation.

Molecular manufacturing requires an error rate that is extremely low by most standards, but is quite permissive compared to the error rates necessary for digital computers. Error rate is an extremely important topic, and unfortunately, understanding of errors in mechanically guided chemistry is susceptible to incorrect intuitions from chemistry, manufacturing, and even physics (many physicists assume that entropy must increase without considering that the system is not closed). It appears that the nature of covalent bonds provides an automatic error-reducing mechanism that will make molecular manufacturing closer in significant ways to computer logic than to today's manufacturing or chemistry.

Recycling Nano-Products

We are often asked, "How will nanofactory-built products be recycled?"

One of the advantages of molecular manufacturing is that it will use very strong and high-performance materials. Most of them probably will not be biodegradable. So what will save us from being buried in nano-litter?

The first piece of good news is that nano-built products will use materials more efficiently. Mechanical parts can be built mostly without defects, making them a lot stronger than today's materials. Active components can be even more compact, because scaling laws are advantageous to small machines: motors may have a million times the power density, and computers may be a million times as compact. So for equivalent functionality, nano-built products will use perhaps 100-1000 times less material. In fact, some products may be so light that they have to be ballasted with water. (This would also make carbon-based products fireproof.)

The second good news is that carbon-based products will burn once any water ballast is removed. Traditionally, incineration has been a dirty way to dispose of trash; heavy metals, chlorine compounds, and other nasty stuff goes up the smokestack and pollutes wherever the wind blows. Fortunately, one of the first products of molecular manufacturing will be efficient molecular sorting systems. It will be possible to remove the harmless and useful gases from the combustion products—perhaps using them to build the next products—and send the rest back to be re-burned.

The third good news is that fewer exotic materials and elements should be needed. Today's products use a lot of different substances for different jobs. Molecular manufacturing, by contrast, will be able to implement different functions by building different molecular shapes out of a much smaller set of materials. For example, carbon can be either an insulator or a conductor, shapes built of carbon can be both flexible or rigid, and carbon molecules can be transparent (diamond) or opaque (graphite).

Finally, it may be possible to build many full-size products out of modular building blocks: microscopic nanoblocks that might contain a billion atoms and provide flexible functionality. In theory, rather than discarding and recycling a product, it could be pulled apart into its constituent blocks, which could then be reassembled into a new product. However, this may be impractical, since the nanoblocks would have to be carefully cleaned in order to fit togethe precisely enough to make a reliable product. But re-using rather than scrappin products is certainly a possibility that's worth investigating further.

Not surprisingly, there is also some bad news. The first bad news is tha carbon is not the only possible material for molecular manufacturing. It i probably the most flexible, but others have been proposed. For example sapphire (corundum, alumina) is a very strong crystal of aluminum oxide. I will not burn, and alumina products probably will have to be scrapped int landfills if their nanoblocks cannot be re-used. Of course, if we are still usin industrial abrasives, old nano-built products might simply be crushed an used for that purpose.

The second bad news is that nano-built products will come in a range o sizes, and some will be small enough that they will be easy to lose. Let m stress that a nano-built product is not a grey goo robot, any more than a toaste is. Tiny products may be sensors, computer nodes, or medical devices, but the will have specialised functionality—not general-purpose manufacturin capability. A lost product will likely be totally inert. But enough tiny los products could add up to an irritating dust.

The third bad news is that widespread use of personal nanofactories wil make it very easy and inexpensive to build stuff. Although each product wil use far less material than today's versions, we may be using far more products

Some readers may be wondering about "disassemblers" and whether the could be used for recycling. Unfortunately, the "disassembler" described in Eri Drexler's *Engines of Creation* was a slow and energy-intensive research tool not an efficient way of taking apart large amounts of matter. It might be possible to take apart old nano-products molecule by molecule, but it would probably be less efficient than incinerating them.

Collecting products for disposal of is an interesting problem. Large products can be handled one at a time. Small and medium-sized products might be enough of a hassle to keep track of that people will be tempted to use them and forget them. For example, networked sensors with one-year batteries might be scattered around, used for two months, and then forgotten—better models would have been developed long before the battery would wear out. In such cases, the products might need to be collected robotically. Any product big enough to have an RFID antenna would be able to be interrogated as to its age and when it was last used. Ideally, it would also tell who its owner had been, so the owner could be billed, fined, or warned as appropriate.

This essay has described what could be. Environmentally friendly cleanup and disposal schemes will not be difficult to implement in most cases. However, as with so much else about molecular manufacturing, the availability of good choices does not mean that the best options necessarily will be chosen. It is likely that profligate manufacturing and bad design will lead to some amount of nano-litter. But the world will be very fortunate if nano-litter turns out to be the biggest problem created by molecular manufacturing.

New Opportunities for DNA Design

DNA is a very versatile molecule, if you know how to use it. Of course, the genetic material for all organisms (except some viruses) is made of DNA. But it is also useful for building shapes and structures, and it is this use that is most interesting to a nanotechnologist.

Readers familiar with DNA complementarity should skip this paragraph. Non-technical readers should read my earlier science essay on DNA folding. Briefly, DNA is made of four molecules, abbreviated A, C, G, and T, in a long string (polymer). G and C attract each other, as do A and T. A string with the sequence AACGC will tend to attach itself to another string with the sequence GCGTT (the strings match head-to-tail). Longer sequences attach more permanently. Heating up a mixture of DNA makes the matched strings vibrate apart; slowly cooling a mixture makes the strings reattach in (hopefully) their closest-matched configuration.

Until recently, designing a shape out of DNA was a painstaking process of planning sequences that would match in just the right way – and none of the wrong ways. Over the years, a number of useful design patterns were developed, including ways to attach four strands of DNA side by side for extra stiffness; ways to make structures that would contract or twist when a third strand was added to bridge two strands in the structure; and three-way junctions between strands, useful for building geometric shapes. A new structure or technique would make the news every year or so. In addition to design difficulties, it was hard to make sufficiently long error-free strands to form useful shapes.

A few months ago, a new technique was invented by Dr. Paul Rothemund. Instead of building all the DNA artificially for his shapes, he realised that half of it could be derived from a high-quality natural source with a fixed but random sequence, and the other half could be divided into short, easily synthesized pieces – "staples" – with sequences chosen to match whatever sequence the natural strand happens to have at the place the staple needs to attach. Although the random strand will tend to fold up on itself randomly to some extent, the use of large numbers of precisely-matching staples will pull it into the desired configuration.

If a bit of extra DNA is appended to the end of a staple piece, the extra bit will stick out from the shape. This extra DNA can be used to attach multiple shapes together, or to grab on to a DNA-tagged molecule or particle. This implies that DNA-shape structures can be built that include other molecules for increased strength or stiffness, or useful features such as actuation.

Although the first shapes designed were planar, because planar shapes are easier to scan with atomic force microscopes so as to verify what's been built, the stapling technique can also be used to pull the DNA strand into a three-dimensional shape. So this implies that with a rather small design effort (at least by the standards of a year ago), 3D structures built of DNA can be constructed, with "Velcro" hanging off of them to attach them to other DNA structures, and with other molecules either attached to the surface or embedded in the interior.

The staple strands are short and easy to synthesize (and don't need to be purified), so the cost is quite low. According to page 81 of Rothemund's notes [PDF], a single staple costs about $7.00 – for 80 nmol, or 50 quintillion molecules. Enough different staples to make a DNA shape cost about $1,500 to synthesize. The backbone strand costs about $12.50 per trillion molecules. Now, those trillion molecules only add up to 4 micrograms. Building a human-scale product out of that material would be far too costly. But a computer chip with only 100 million transistors costs a lot more than $12.50.

The goal that's the focus of this essay is combining a lot of these molecular "bricks" to build engineered heterogeneous structures with huge numbers of atoms. In other words, rather than creating simple tilings of a few bricks, stick them together in arbitrary computer-controlled patterns, constructs in which every brick can be different and independently designed.

I was hoping that nano-manipulation robotics had advanced to the point where the molecular shapes could be attached to large handles that would be grabbed and positioned by a robot, making the brick go exactly where it's wanted relative to the growing construct, but I'm told that the state of the art probably isn't there yet. Just one of the many problems is that if you can't sense the molecule as you're positioning it, you don't know if temperature shifts have caused the handle to expand or contract. However, there may be another way to do it.

An atomic force microscope (AFM) uses a small tip. With focused ion beam (FIB) nano-machining, the tip can be hollowed out so as to form a pocket suitable for a brick to nestle in. By depositing DNA Velcro with different sequences at different places in the pocket (which could probably be done by coating the whole tip, burning away a patch with the FIB, then depositing a different sequence), it should be possible to orient the brick relative to the tip. (If the brick has two kinds of Velcro on each face, and the tip only has one kind deposited, the brick will stick less strongly to the tip than to its target position.)

Now, the tip can be used for ordinary microscopy, except that instead of having a silicon point, it will have a corner of the DNA brick. It should still be usable to scan the construct, hopefully with enough resolution to tell where the tip is relative to the construct. This would solve the nano-positioning problem.

I said above that the brick would have DNA Velcro sticking out all over. For convenience, it may be desirable to have a lot of bricks of identical design, floating around the construct – as long as they would not get stuck in places they're not wanted. This would allow the microscope tip to pick up a brick from solution, then deposit it, then pick up another right away, without having to move away to a separate "inkwell." But why don't the bricks stick to the construct and each other, and if they don't, then how can the tip deposit them, and why do they stay where they're put?

To make the bricks attach only when and where they're put requires three conditions. First, the Velcro should be designed to be sticky, so that the bricks will stay firmly in place once attached. Second, the Velcro should be capped with other DNA strands so that the bricks will not attach by accident. Third, the capping strands should be designed so that physically pushing the brick against a surface will weaken the bond between Velcro and cap, allowing the Velcro to get free and bind to the target surface. For example, if the cap strands stick stiffly out away from the block (perhaps by being bound to two Velcro strands at once), then mechanical pressure will weaken the connection.

Mechanical pressure, of course, can be applied by an AFM. Scan with low force, and when the brick is in the right place, press down with the microscope. Wait for the cap strands to float away and the Velcro to pair up, and the brick is deposited. With multiple Velcro strands between each brick, the chance of them all coming loose at once and allowing the brick to be re-capped can be made miniscule, especially since the effective concentration of Velcro strands would be far higher than the concentration of cap strands.

But before the brick was pushed into place, the chance of all the cap strands coming loose at once also would have been miniscule. (For any physicists reading this, thermodynamic equilibrium between bound and free bricks still applies, but the transition rate can be made even slower than the above concentration argument implies, since the use of mechanical force allows an extremely high energy barrier. If the mechanical force applied is 100 pN over 5 nm, that is 500 zJ, approximately the dissociation energy of a C-C bond.)

So, it seems that with lots of R&D (but without a whole lot of DNA design), it might be possible to stick DNA bricks (plus attached molecules) together in arbitrary patterns, using an AFM. But an AFM is going to be pretty slow. It would be nice to make the work go faster by doing it in parallel. My NIAC project suggests a way to do that.

The plan is to build an array of "towers" or "needles" out of DNA bricks. In the tip of each one, put a brick-binding cavity. Use an AFM to build the first one in the middle of a flat surface. Then use that to build a second and third needle on an opposing surface. (One of the surfaces would be attached to a nano-positioner.) Use those two towers to build a fourth, fifth, sixth, and seventh on the first surface. The number of towers could grow exponentially.

By the time this is working, there may be molecules available that can act as fast, independently addressable, nanoscale actuators. Build a mechanism into each tower that lets it extend or retract – just a nanometre or so. Now, when the towers are used to build something, the user can select which bricks to place and which ones to hold back. This means that different towers, all attached to the same surface and moved by the same nano-positioner, can be doing different things. Now, instead of an exponential number of identical designs, it has become possible to build an exponential number of different designs, or to work on many areas of a large heterogeneous design in parallel.

A cubic micron is not large by human standards, but it is bigger than most bacteria. There would be about 35,000 DNA bricks in a cubic micron. If a brick could be placed every fifteen seconds, then it would take a week to build a cubic micron out of bricks. This seems a little fast for a single AFM that has to bind bricks from solution, find a position, and then push the brick into place, even if all steps were fully automated – but it might be doable with a parallel array (either an array of DNA needles, or a multi-tip AFM). If every brick were different, it would cost about $50 million for the staples, but of course not every brick will be different. For 1,000 different bricks, it would cost only about $1.5 million.

We will want the system to deposit any of a number of brick types in any location. How to select one of numerous types? The simplest way is to make all bricks bind to the same tip, then flush them through one type at a time. This is slow and wasteful. Better to include several tips in one machine, and then flush through a mixture of bricks that will each bind to only one tip. The best answer, once really high-function bricks are available and you're using DNA-built tips instead of hollowed-out AFM tips, is to make the tips reconfigurable by using fast internal actuators to present various combinations of DNA strands for binding of various differently-tagged bricks.

I started by suggesting that a scanning probe microscope be used to build the first construct. Self-assembly also could be used to build small constructs, if you can generate enough distinct blocks. But you may not have to build hundreds of different bricks to make them join in arbitrary patterns. Instead, build identical bricks, and cap the Velcro strands with a second-level "Velcro staple." Start with a generic brick coated with Velcro – optionally, put a different Velcro sequence on each side. Stir that together with strands that are complementary to the Velcro at one end and contain a recognition pattern on the other end. Now, with one generic brick and six custom-made Velcro staples, you have a brick with a completely unique recognition pattern on each side. Do that for a number of bricks, and you can make them bind together any way you want.

One possible problem with this is that DNA binding operations usually need to be "annealed" – heated to a temperature where the DNA falls apart, then cooled slowly. This implies that the Velcro-staple approach would need three different temperature ranges: one to form the shapes, one to attach the staples, and one to let the shapes join together.

The Velcro-staple idea might even be tested today, using only the basic DNA-shape technology, with one low-cost shape and a few dozen very-low-cost staples. Plus, of course, whatever analysis tools you'll need to convince you that you're making what you think you're making.

There is a major issue involved here that I have not yet touched on. Although the DNA staple technique makes a high percentage of good shapes, it also makes a lot of broken or incomplete shapes. How can the usable shapes be sorted from the broken shapes? Incomplete shapes may be sorted out by chromatography. Broken shapes might possibly be rejected by adding a fluorescence pair and using a cell sorter to reject shapes that did not fluoresce properly. Another possibility, if using a scanning probe microscope (as opposed to the "blind" multi-needle approach) is to detect the overall shape of the brick by deconvolving it against a known surface feature, and if an unwanted shape is found, heat the tip to make it dissociate.

This is just a sketch of some preliminary ideas. But it does go to show that the new DNA staple technology makes things seem plausible that would not have been thinkable before it was developed.

Military Implications of Molecular Manufacturing

This essay will survey the technology of molecular manufacturing, the basic capabilities of its products, some possible weapon systems, some tactical and strategic considerations, and some possible effects of molecular manufacturing on the broader context of societies and nations. However, all of this discussion must take place in the context of the underlying fact that the effects and outcome of molecular manufacturing will be almost inconceivable, and certainly not susceptible to shallow or linear analysis.

Take a minute and try to imagine a modern battlefield without electricity. No radar or radios; no satellites; no computers; no night vision, or even flashlights; no airplanes, and few ground vehicles of any kind. Imagination is not sufficient to generate this picture—it simply doesn't make sense to talk of a modern military without electricity.

Molecular manufacturing will have a similarly profound effect on near-future military affairs.

Electricity is a general-purpose energy technology, useful for applications from motors to data processing. A few inventions, ramified and combined—the storage battery, transistor, electromagnet, and a few others—are powerful enough to be necessary components of almost all modern military equipment and activities.

If it is impossible to conceive of a modern military without electricity—a technology that exists, and the use of which we can study—it will be even less feasible to try to imagine a military with molecular manufacturing.

Molecular manufacturing will be the world's first general-purpose manufacturing technology. Its products will be many times more plentiful, more intricate, and higher performance than any existing product. They will be built faster and less expensively, speeding research and development. They will cover a far greater range of size, energy, and distance than today's weapons systems.

As increasingly powerful weapons make the battlefield untenable for human soldiers, computers vastly more powerful and compact than today's will enable far higher degrees of automation and remote operation. Kilogram-scale manufacturing systems, building directly from the latest blueprints in minutes, will utterly transform supply, logistics, and deployment.

Radium and X-rays were discovered within months of each other. Within a few years, X-rays had inspired stories about military uses of "death rays." Decades later, Madame Curie gave speeches on the wonderful anti-cancer properties of radium. It would have been difficult or impossible to predict that a few decades after that, X-rays would be a ubiquitous medical technology, and nuclear radiation would be the basis of the world's most horrific weapons. While reading the rest of this article, keep in mind that the implications of various molecular manufacturing products and capabilities will be at least as unpredictable and counterintuitive.

Risks from Military Uses of Nanotechnology

Promises versus Risks

There is a tendency in the world of high technology to over-advertise conceivable benefits; researchers, funding officers and entrepreneurs understand that investors, politicians and the public have come to expect hype and will react with a yawn to any technical proposal that does not promise to cure cancer, provide limitless energy, prevent terrorism or make stupid people smart. But speak of the unproven hazards posed by future technologies, and one is fast denounced as a doomsayer, a purveyor of science fiction, guilty of exaggeration and extrapolation from dubious premises. In the field of nanotechnology, warnings have been sounded against excessive promises made too soon, lest the public become disillusioned with "nanohype," but the roughest criticism has been dealt to those who not only foretold bountiful results from this new technology, but also warned of grave dangers.

Such warnings, many believe, will lead the public to exaggerated fears of the unknown, undercutting support for nanotechnology funding. We believe it is essential to get beyond this battle over atmospherics, and to commence balanced and careful scholarship to assess the actual prospects for good or evil, and to consider what should be done in response to them. We write to address the question of dangers arising from the military use of nanotechnology.

Risks of Nanotechnology

Nanotechnology (NT) holds great promises, but also poses grave risks. This applies even when considering only the evolutionary advances expected from current laboratory research and generally-accepted extrapolations of historical trends. It is strongly apparent in the context of visions such as nanoassemblers, self replication, artificial intelligence of human capability and beyond, robotics from nano to macroscale, super-automated production, and nanodevices within the human body, perhaps to eradicate illness, perhaps to interact with the brain. Although there are disputes about the realisibility of these latter concepts, caution demands serious consideration of such prospects unless they can be shown to be physically or technically impossible. To prevent irreversible damages, regulatory measures must be taken in advance of dangerous developments.

The risks of NT span a wide range: environmental pollution, increase of inequality, displacement of human workers or even of the human species have been mentioned. Interdisciplinary studies should be undertaken to address these various risks. Here, we want to draw attention to risks linked to military NT activities that could create specific dangers as well as accelerate general developments in such a way that in-depth study and informed decision become more difficult. Military exploitation of NT has barely begun, but there are strong indications it may expand rapidly, driving and in turn being driven by the technology. Given that the US National Nanotechnology Initiative (NNI) has stimulated similar initiatives in many other countries, the USA may also provide a role model for military R&D there.

Potential Military NT Uses

Weapons of mass destruction. Self-replicating nanorobots, aggressively consuming organic material, are perhaps the most oft-mentioned, and perhaps overstated concept, but would probably require an advanced stage of NT development.

In the nearer term, NT will provide possibilities in coming decades for more efficient storage, dispersal, and transport of chemical and biological agents into the body and cells of humans, animals, or plants. New agents may remove previous operational difficulties of biological warfare. Advanced capabilities may include the use of genetic markers to target an ethnic group or even a specific individual. New options for nuclear weapons might include NT-based materials extraction and processing, weapons production, and perhaps new types of nuclear weapons. NT manufacturing based on self-replication could produce conventional weapons in such large quantities that they acquire the character of mass-destruction weapons.

Other weapons. NT will provide stronger, lighter materials, smaller computer components, new sensor technologies, and-together with and beyond microsystems technologies-many options for miniaturization. NT manufacturing methods may enable mass production of sophisticated expendable systems at low cost. One can foresee greater projectile velocities, stronger light armor, and precision-guidance systems even in small munitions. Low-cost military robots of mini and micro sizes, including biological-technical hybrids, would bring even more radical changes. Such robots would necessarily be capable of autonomous decision and action and could be used for purposes from reconnaissance to attack. In outer space, very small satellites could act as anti-satellite weapons.

General military applications. NT will have applications in energy storage and generation, propulsion, displays, sensors and sensor nets, combat information systems, logistics, maintenance, self-repair, smart materials, and more. Some more visionary concepts foresee systems implanted into soldiers' bodies, first for biomedical analysis and reaction, later for information exchange.

Far-Fetched Fears

> *"Third, social and cultural biases to a brain implanted decision tool must be overcome. The Cyber Situation is designed to assist, not control each decision maker. To fully exploit growing technology, cumbersome hardware and software requirements must be reduced to the simplicity and seamlessness of a chip implant."*
>
> *"Nanotechnology will allow the miniaturisation of sensors and equipment.... Postulated systems, some further out than 2030, include nano-solar cells offering more efficient electricity generation than present systems, and nano-robots for many potential purposes, but including medical robots that can fight diseases internally in humans, remove and replace defective DNA, and even possibly treat injured personnel. Combined with*

increases in processing power these systems will have widespread application including in mini-platforms for reconnaissance."

"Beyond technological obstacles, the potential for effective battlefield robots raises a whole series of strategic, operational, and ethical issues, particularly when or if robots change from being lifters to killers. The idea of a killing system without direct human control is frightening.... Should the United States attempt to control the proliferation of military robotic technology? Is that even feasible since most of the evolution of robotic technology, like information technology in general, will take place in the private sector? Should a fully roboticized force be the ultimate objective?"

Dangers and Risks of Military NT Uses

When taking a preliminary look at NT under the criteria of preventive arms control, several dangers come to mind, in all three problem areas. These concern: - arms control agreements (e.g., Biological Weapons Convention through new NT-genetics-based agents, limits on conventional forces by new weapons types outside of treaty definitions) or the international law of warfare (e.g., through introduction of autonomous fighting systems not reliably recognising non-combatants or combatants hors de combat), stability (arms races from technological innovation, pressure for preventive attack and fast action, proliferation of cheap microsystems), humans, the environment, or society (microrobots for eavesdropping, crime, and terrorism; uncontrolled self-replication; implanted systems altering human nature).

In the third category dealing with peacetime civilian life, military research and deployment of systems could create "facts" before society is able carry out a thorough debate about the desirability of particular technological developments.

Preventive Arms Control-A Concept for Limiting Risks from New Military Technology

Criteria of Preventive Arms Control:

(1) dangers to arms control agreements and the international law of warfare,

(2) dangers to stability (first strike, arms race, and proliferation)

(3) dangers to humans, environment, or society.

Steps of Preventive Arms Control:

(1) prospective scientific analysis of the technology,

(2) prospective analysis of the military-operational aspects,

(3) assessment of both under the criteria,

(4) devising possible limits and verification methods.

Regulation of Technology-Precedents:

National Level: Laws and standards for environment, occupational safety, research (e.g., genetics) - with strong inspections

International Level, General: UN Convention on the Law of the Sea (1982, pollution, exploitation, technology cooperation), UN Framework Convention on Climate Change (1992, reduce greenhouse-gas emissions)

International Level, Arms Control: Anti-Ballistic Missile Treaty (1972, no deployment, testing, and development of space-based anti-ballistic missile systems) (now terminated by the USA); Biological and Toxin Weapons Convention (1972, no possession, production, and development of agents for non-peaceful purposes); Blinding Laser Weapons Protocol (1995, bans use of laser weapons designed to produce permanent blindness).

Recommendations

This purpose of this paper is to raise the international awareness of the dangers that could ensue from military NT activities. Detailed, interdisciplinary studies of these risks are needed, and preventive-arms-control measures should be worked out. Because military and civilian aspects are intertwined, limitations and rules should encompass both, of course duly considering also the beneficial NT uses.

At present we propose the following general guidelines:

- No circumvention of existing treaties.
- A comprehensive ban on space weapons.
- Prohibition of autonomous "killer robots."
- Specific restrictions on small autonomous systems.

As first steps, we suggest:

- Measures to prevent production or release of systems capable of self-replication in the wild should be concluded internationally, binding both the civilian and military sectors.

The nanotechnology initiatives of various nations should work together to build confidence and common purpose, and to address concerns such as arms control, safety protocols, and social implications.

In the long run, containing the risks of the new powerful technologies-genetic engineering, pervasive computer networks, micro-systems, and nanotechnology-will probably require fundamental changes in the international system, particularly strengthening of law and political institutions, including international criminal law, and reorienting of the military mission from warfighting to organising cooperative security.

Areas of Application for Nanotechnology and Nanoscience

Nanotechnology is known as the technology of the 21st century. At the present time, many applications are already known, including data storage, restoring data (Giant Magneto-Resistive (GMR) heads for hard disc drives in computers), sun creams, airbag sensors and scratch-resistant coatings. The limits of future applications and areas of application are far from being reached. On NanoNed it is stated that nanoscience research generates possibilities for creating better and cheaper products for daily life with the aid of sustainable production processes. The following areas of application for nanotechnology can be identified:

- Nanomaterials/Industrial production
- Nano-electronics
- Bio-electronics
- Nanotechnology in medicine
- Military technology
- Agricultural nanotechnology

Other areas of application, including applications in the energy sector and consumer products We will be illustrating four areas of application in detail below: nano-electronics, nanotechnology in medicine, agricultural nanotechnology and defence. Information on all areas of application can be found in the outline *Om het kleine te waarderen* [To value small things] by the Rathenau Institute (2004), written in Dutch.

5

The Threshold of Nanotechnology

In the previous chapter, we looked at the state of current research, but from there to the nanotechnology of even the Pocket Library scenario is a leap. How will this gap be crossed?

In this chapter, we outline how emerging technologies can lead to nanotechnology. The actual path to nanotechnology-the one that history books will record-could emerge from any one of the research directions in physics, biochemistry, and chemistry recounted in the last chapter, or (more likely) from a combination of them.

The availability of so many good options builds confidence that the goal can be reached, even while it decreases confidence that some particular path will be fastest. To see how advances might cross the gap from present technology to early nanotechnology, let's follow one path out of the many possible.

Bridging the Gap

One way to bridge the gap would through the development of an AFM-based molecular manipulator capable of doing primitive molecular manufacturing. This device would combine a simple molecular device-a molecular gripper-with an AFM positioning mechanism. An AFM can move its tip with precision; a molecular manipulator would add a gripper to the tip to hold a molecular tool. A molecular manipulator of this kind would guide chemical reactions by positioning molecules, like a slow, simple, but enormous assembler. (In our standard simulation view, where a molecular assembler arm fits in a room, the AFM apparatus of a molecular manipulator would be the size of a moon.) Despite its limits, an AFM molecular manipulator will be a striking advance.

How might this advance occur? Since we're choosing one path out of many possible, we may as well include more details and tell a story. (A more technical description of a device like the following can be found in Nature).

Developing a Molecular Manipulator

Several years ago, researchers at the University of Brobdingnag began work on developing a molecular manipulator. To reach this goal, a team of a dozen physicists, chemists, and protein researchers banded together (some working full time, some part time) and began the creative teamwork needed to solve the basic problems.

First they needed to attach a gripper to an AFM tip. As grippers, they chose fragments of antibody molecules, the selectively sticky proteins that the immune system uses to bind and identify germs. If they could get the "back" of the molecule stuck onto a tip, then the "front" could bind and hold molecular tools. (The advantage of antibody fragments was this: freedom of tool choice. Since the late 1980s, researchers had been able to generate antibodies able to bind almost any preselected molecule-or molecular tool.)

A molecular manipulator (AFM tip and tool holder, above) would bind and position reactive molecular tools to build up a workpiece, molecule by molecule.

In parallel, the U. Brob AFM researchers worked on placing tips in a precise location and then holding them there with atomic accuracy for seconds at a time. This proved straightforward. They used techniques developed elsewhere during the early 1990s, adding only modest refinements.

They now had their gripper and a way of putting it where they wanted it, but they needed a set of tools. The gripper was like the chuck of a drill, waiting to have different bits fitted into its tool-holder slot. So as the final step, the synthetic chemists on the team made a dozen different molecular tools, all identical at one end but different at the other.

The similar parts all bound to the same antibody tool-holder, slotting neatly into position. The different parts were all chemically reactive in different ways. Like the molecular tools in the hall of assembler arms, each of these tools could use a chemical reaction to transfer some atoms to a molecular object under construction.

Developing the molecular tool kit was the toughest part of the project; it took about as much work as had gone into duplicating the palytoxin molecule back in the 1980s. None of the tasks in the project demanded the solution of a deep scientific puzzle, and none demanded the solution of a notoriously difficult engineering problem. Each task had many possible solutions, the problem was to find a compatible set of solutions and apply them. After a few years, the solutions came together and the U. Brob research team began building new molecules by molecular manipulation. Now many teams are doing likewise.

Molecular Grippers and Tools

To build something with the U. Brob team's AFM-based molecular manipulator system, you use it as follows: First, choose a surface to build on and place it under the tip in a pool of liquid. Then dunk the AFM tip into the liquid, bringing it down to the surface, and back it off a little. Construction can now begin as soon as a tool is loaded into the gripper.

Tubes and pumps can flow different liquids over the surface and past the gripper, carrying different tool molecules. If you want to do something with a tool of Type A, you wash in the proper liquid, and a Type A molecule promptly sticks the to the gripper as shown in Figure. Once it is in the gripper, you can use the AFM mechanism to move it around and put it where you want it. Move it up to the surface at a convenient spot, wait a few seconds, and it reacts, forming a bond and leaving a molecular fragment attached to the spot you chose. To add a different fragment, you can use a tool of Type B: you back up the tip, flow in a fresh liquid carrying the new tools, and in a moment a tool of the new type is bound in place and ready to apply, either on or alongside the first spot. Step by step, you build up a precise molecular structure.

Each step takes only seconds. Molecular tools pop into the gripper in a fraction of a second, and used tools pop off at the same rate. Once the tip has positioned a molecule, it reacts quickly, about a million times faster than unwanted reactions at other sites. In this way, the molecular manipulator gives good control of where reactions will occur (though it is not as reliable as an advanced assembler would be).

It is fairly fast by a chemist's standards-per cycle-but still a million times slower than an advanced assembler. It can perform a variety of steps, but isn't as flexible and capable as an advanced assembler. In short, it is hardly the last word in nanotechnology, yet is a great advance over what has gone before.

Products

With its ability to accelerate desired reactions by a factor of a million or so, the U. Brob team's molecular manipulator can perform 10,000 to 100,000 steps with good reliability. Back in the 1980s, chemists making protein molecules struggled to perform just one hundred steps.

The U. Brob research team (and its many imitators) can now build structures that are stronger and easier to design than proteins: not floppy, folded chains, but rugged objects held together by a sturdy network of bonds. Though not as strong and dense as diamond, these structures are like bits of a tough engineering plastic. A specially adapted computer-aided design system makes it easy to design molecular objects made from these materials.

Yet the AFM-based molecular manipulator has one grave disadvantage: It does chemistry one molecule at a time, and it ties up a machine as expensive as a car for hours or days to produce that one large molecule. Some molecules, though, are valuable enough to be worth building even one at a time. These draw prompt attention.

A single molecule isn't much use as a dye, a drug, or a floor wax, but it can have substantial value if it provides useful information. The U. Brob team quickly publishes a pile of scientific papers based on experiments with single molecules: they build a molecule, probe it, report the results, and build another. Some of these results show chemists elsewhere in the multibillion-dollar chemical industry how to design new catalysts, molecules that can help make other molecules more cheaply, cleanly, and efficiently. This information is worth a lot.

Three new products of special interest are among the first to be made. The first-molecular electronics-begins with experiments conducted by a research group at a computer chip company. They use their molecular manipulator to build single molecules and probe them, gradually learning how to build the parts needed for molecular electronic computers. These new computers don't immediately become practical, because the costs are too high for making such large molecules with AFM-based technology. Yet some companies begin to produce simpler molecular electronic devices for use in sensors and specialised high-speed signal processing. A speciality industry is born and begins to expand.

The second product is a gene reader, a complex molecular device built on the surface of a chip. The biologists who built the reader combined proteins borrowed from cells with special-purpose molecular machines designed from scratch. The result was a molecular system that binds DNA molecules and pulls them past a read-head-like tape through a tape recorder. The device works as fast as some naturally occurring molecular machines that read DNA, with one key advantage: it outputs its data electronically. At that speed, a single device can read a human genome in about a year. Though still too expensive for a doctor's office, these readers are promptly in great demand from research laboratories. Another small industry is born.

The third product is far more important, in the long run: replacement tips for molecular manipulators, grippers, and tools that are better than the originals. With these new, more versatile devices, researchers are now building more ambitious products and tools.

The Next Step to Nanotechnology

While the physicist-led team at U. Brob was finishing its work on the AFM-based molecular manipulator, a chemist-led team at the University of Lilliput was working furiously. They saw the U. Brob desktop machine as too large and its expected products as too expensive.

Even back in the 1980s, David Biegelsen of the Xerox Palo Alto Research Centre had noted, "The main drawback I see to using a hybrid protoassembler [AFM-based molecular manipulator] is that it would take a long time to build just one unit. Building requires a series of atom-by-atom construction steps It would be better to build in parallel from the very beginning, making many trillions of these molecules all at the same time.

I think there is tremendous power in parallel assembly. Maybe another field, chemistry or biology, offers a better way to do it." The chemists at U Lill aimed to develop that better way, building first simple and then more and more complex molecular machines. The eventual result was a primitive molecular assembler able to build molecular objects by the trillions.

Chemist's Tools

How did the chemists achieve this? During the years when the U. Brob team was developing the molecular manipulator, researchers working in protein science and synthetic chemistry had made better and better systems of molecular building blocks. Chemists were well prepared for doing this: by the late 1980s, it had become possible to build stable structures the size of medium-sized protein molecules, and work had begun to focus on making these molecules perform useful work by binding and modifying other molecules.

Chemists learned to use these sophisticated catalysts-early molecular devices-to make their own work easier by helping in the manufacture of still more large molecules. Another traditional chemist's tool was software for doing computer-aided design. The early software designed by Jay Ponder and Frederic Richards of Yale University ultimately led to semi-automatic tools for designing molecules of a particular shape and function. Chemists then could easily design molecules that would self-assemble into larger structures, several tens of nanometres across.

Molecular Construction Machines

These advances in software and chemical synthesis let the U. Lill team tackle the task of building a primitive version of a molecular assembler. Although they couldn't build anything as complex as a nanocomputer or as stiff as diamond, they didn't need to. Their design used sliding molecular rods to position a molecular gripper much like the gripper used at U. Brob, again using the surrounding liquid to control which tool the gripper held. Instead of an AFM's electronic controls, they used the surrounding liquid to control the position of the rods as well. In a neutral solution, the rods would withdraw; in an acid solution, they would extend. How far they moved depended on what other molecules were around to lodge in special pockets and block the motion.

Their primitive assemblers built much the same sorts of products that the U. Brob molecular manipulator did; the tools were similar, and speed and accuracy were about the same. Yet there was one dramatic advantage: About 1,000,000,000,000,000,000,000 U. Lill assemblers could fit in the space occupied by one U. Brob manipulator, and it was easy to produce a mere 1,000,000,000,000,000 times as much product at the same cost.

With the first, primitive assemblers, construction was slow because each step required new liquid baths and several seconds of soaking and waiting, and a typical product might take thousands of steps. Nonetheless, the U. Lill team made a lot of money licensing their technology to researchers trying to commercialize products they had first researched with the U. Brob machine.

After starting an independent company (Nanofabricators, Inc.), they poured their research efforts into building better machines. Within a few years, they had assemblers with multiple grippers, each loaded with a different kind of tool; flashes of coloured light would flip molecules from state to state (they copied these molecules from the pigments of the retina of the eye); flipping molecules would change tools and change rod positions. Soaking and waiting become a thing of the past, and soon they were pouring out parts that, when mixed with liquid and added to dishes with special blank chips would build up the dense memory layers that made possible the Pocket Library.

That was when things started moving fast. The semiconductor industry went the way of the vacuum tube industry. Money and talent poured into the new technology. Molecular CAD tools got better, assemblers made it easy to build what was designed, and fast production and testing made molecular engineering as easy as playing with software. Assemblers got better, faster, and cheaper.

Researchers used assemblers to build nano-computers, and nanocomputers to control better, faster assemblers. Using tools to build better tools is an ancient story. Within a decade, almost anything could be made by molecular manufacturing, and was.

Will developments in the late pre-breakthrough days be as just described? Certainly not: the technical approaches will differ, and the U.S. academic research setting implied by the scenario could easily be replaced by academic, commercial, governmental, or military research settings in any of the advanced nations. What do seem realistic are the implied requirements for effort, technology, and time, as well as the basic capabilities of different devices. We are approaching a threshold of capability beyond which further advances will become easy and fast.

Working with the Nanotechnology

The word manufacturing comes from the Latin manufactus, meaning "handmade." Today, the term brings to mind huge, noisy machines stamping out products and spewing waste. Giving up manufactured products isn't popular or practical-almost everything we use today is manufactured. If all machine-made products were to suddenly vanish, most people in the United States would find themselves naked and outdoors, with very little around them. Expanding manufacturing is an objective of nearly every nation on Earth.

We can't give up manufacturing, but we can replace today's technologies with something radically different. Molecular manufacturing can help us get what we seem to want: high-quality products made at low cost with little environmental impact. Chapter 12 will describe the grave problems raised by misapplication of this capability, but for now we discuss the positive side.

What follows is an exploration of the possible-a look at the devices that could be built once precise molecular control is achieved, and a look at how people might run a manufacturing business based on nanomachines.

Try not to think of these sketches as hard-and-fast predictions of precisely how things will be done, but instead as descriptions of capabilities-the sorts of things that can be done once nanotechnology is well in hand. Doubtless there will be better ways to do things than the ways we describe. As usual, references to the 1980s and before are historically accurate; in the rest, the science isn't fiction..

The Desert Rose Industries

Desert Rose Industries is a diversified wholesale manufacturer of enough furniture, computers, toys, and recreation equipment to have made any twentieth-century captain of industry proud. But if you assembled all Desert Rose employees in front of corporate headquarters, you'd see Carl and Maria Santos standing beside a building the size of a four-bedroom house.

This industrial giant is a typical mom-and-pop business, helped along by a network of telecommuters who handle sales and customer support from homes scattered across North America.

Their friends chide Carl and Maria as "old-fashioned traditionalists" and tease Maria about abandoning Carl in the factory while she travels to Europe, Asia, South America, and Africa for new business. In the molecular-manufacturing business, familiar personal skills and virtues-honesty, accuracy, good communication-are as important as before. Maria likes to work with the customers.

Aided by her S.B. in molecular manufacturing from MIT and her MFA in design, she patiently helps nervous new designers through their first manufacturing experience, and with unflagging courtesy and good humor, handles rush orders, last-second changes, and special orders. Maria's good design ideas and caring personality won them a reputation for being responsive to customer needs. Carl, precise and careful, built their name for accurate manufacturing and delivery on schedule.

Except for Carl's habit of playing Gershwin at full volume with the windows open, the only sounds at the Desert Rose site are the birds along the banks of the stream that winds across the canyon floor; no clanking machinery here. Maria's parents built Desert Rose Industries out here on an old smelter many miles away from human neighbours. They regraded the land and cleaned up the wastes. Maria adapted a molecular processor to convert heavy-metal contaminants back into stable minerals, and shipped them off to help refill the hole they had originally come from, an old open-pit mine. The desert has mostly healed now, and a few tough trees are spreading along the stream again.

New customers coming up the road for a firsthand look at the manufacturing operations get the full tour: a lunch/meeting room, Maria's office, the manufacturing plant, and the warehouse space for parts and products out back. "The plant" is the largest room, and Carl's pride. Twelve manufacturing ponds and their cooling systems-vats ranging in size from a kitchen sink to a small swimming pool-are where Desert Rose uses nanocomputers and assemblers to do their building work.

A plumbers' nightmare of piping runs between the ponds and a triple row of containers with labels like CARBON FEEDSTOCK, PREPARED PLATINUM, SIZE-4 STRUC-TURAL FIBRES, and PREFAB MOTORS. Carl keeps a good stock of parts and raw materials on hand, with more in the underground warehouse. Sure, some rare things almost never get used, but having them ready to go is one of Carl's secrets for delivering on time and building precisely to specification.

Over on a table are Carl's music system and the computers-descendants of the IBM PCs and Macintoshes of the 1980s-that are used to run the manufacturing process. In a space the size of a large living room, Carl and Maria have all the raw materials and all the production equipment-nanocomputers and assemblers-they need for building almost anything.

Occasionally, Carl and Maria need the services of specialised tools, such as disassemblers, that might exist only in labs. A disassembler works like an archeologist, painstakingly excavating the structure of a molecule, removing atom after atom, in order to record and analyze the molecular structure.

Because they work so slowly, noting the position of each molecule, disassemblers aren't used for recycling operations-it would be expensive and pointless to record all this unwanted data. But as tools for analyzing the unknown, they're hard to beat..

Maria found this out when a customer sent her an order for tropically scented furniture and fixtures for his restaurant, but instead of including the software instructions for building the perfume, Maria found a plastic bag full of resinous brown gook with a note saying, "I got this stuff in the tropics. Please make the fabric smell like this." Maria (after sniffing the gook and deciding it smelled surprisingly tropically good) shipped the sample to the lab for chemical analysis by disassembler. The lab sent back software with the molecular description and instructions for building the same scent into the furniture.

Carl usually schedules production very tightly: in every tank, assemblers are building products; every computer is directing work. But this morning, listening to the tone of Maria's voice wafting in from the front office, Carl changes his plans: something important is about to happen. He postpones building orders for video wallpaper and commemorative diamond baseballs, and holds three pools and a computer ready. Minutes later, Maria hurries in, her voice tight and anxious. "Carl, that earthquake down south-they need help. Amanda from the Red Cross is sending the software right now."

To build a product, Desert Rose needs design instructions-computer software-for the assemblers. Carl and Maria have their own software library, but usually they buy or rent what they need, or the customers send their own designs.

The software that Amanda sends contains the specifications to manufacture the emergency equipment: a set of instructions to be run on a standard desktop computer. Within minutes, two copies of the Red Cross software arrive electronically. Before starting the build, Carl meticulously checks to make sure that the master copy and backup copy agree and weren't damaged in transit. If the instructions are complete and correct and properly signed with the Red Cross data stamp, then the desktop computer will communicate these building instructions directly to millions of small computers acting as on-the-job foremen directing the work: nanocomputers.

Nanocomputers

While the first, primitive assemblers were controlled by changing what molecules are in the solution around the device, getting the speed and accuracy wanted for large-scale manufacturing takes real computation. Carl's setup uses a combination of special-purpose molecule processors and general-purpose assemblers, all controlled and orchestrated by nanocomputers.

Computers back in the 1990s used microelectronics. They worked by moving electrical charge back and forth through conducting paths-wires, in effect-using it to block and unblock the flow of charge in other paths. With nanotechnology, computers are built from molecular electronics.

Like the computers of the 1990s, they use electronic signals to weave the patterns of digital logic. Being made of molecular components, though, they are built on a much smaller scale than 1990s computers, and work much faster and more efficiently. On the scale of our simulated molecular world, 1990s computer chips are like landscapes, while nanocomputers are like individual buildings. Carl's desktop PC contains over a trillion nanocomputers, enough to out-compute all the microelectronic computers of the twentieth century put together.

Back in the dark ages of the 1980s, an exploratory engineer proposed that nanocomputers could be mechanical, using sliding rods instead of moving electrons as shown in Figure earlier. These molecular mechanical computers were much easier to design than molecular electronic computers would have been. They were a big help in getting some idea of what nanotechnology could do.

An electronic transistor lets current flow when a negative electric charge is applied and blocks current when a positive charge is applied. The mechanical "transistor" (below) lets the horizontal rod move when the vertical rod is down, and blocks the horizontal rod when the vertical rod is up. Either device can be used to build logic gates and computers. Even back then, it was pretty obvious that mechanical computers would be slower than electronic computers. Carl's molecular electronic PC would have been no great surprise, though nobody knew just how to design one. When nanotechnology actually arrived and people started competing to build the best possible computers, molecular electronics won the technology race. Still, mechanical nanocomputers could have done all the nanocomputing jobs at Desert Rose: ordinary, everyday molecular manufacturing just doesn't demand the last word in computer performance.

For Carl, the millions of nanocomputers in the milky waters of his building ponds are just extensions of machines on his desk, machines there to help him run his business and deliver products to his customers-or, in the case of the Red Cross emergency, to help provide time-critical emergency supplies. By reserving those three separate ponds, Carl can either build three different kinds of equipment for the Red Cross or use all the ponds to mass-produce the first thing on the Red Cross list: emergency shelters for ten thousand people. The software is ready, the plumbing is fine, the drums of building materials are all topped up, the Special Mix for this job is loaded: the build is ready to start. "Okay," Carl tells the computer, "build Red Cross tents." Computer talks to nanocomputers. In all three pools, nanocomputers talk to assemblers. The build begins.

Products Assembling

Some of the building done at Desert Rose Industries uses assemblers much like the ones we saw in the first hall of the plant tour, back in the simulated molecular world of the Silicon Valley Faire. As seen in simulation, they are big, slow, computer-controlled things moving molecular tools. With the right instructions and machinery to keep them supplied with molecular tools, these general-purpose assemblers can build almost anything.

They're slow, though, and take a lot of energy to run. Some of the building uses special-purpose assembly systems in the molecule-processing style, like the systems in the basement we saw in the tour of a simulated molecular factory. The special-purpose systems are all moving belts and rollers, but no arms. This is faster and more efficient, but for quantity orders, cooling requirements limit the speed.

It's faster to use larger, prefabricated building blocks. Desert Rose uses these for most of their work, and especially for rush orders like the one Carl just set up. Their underground warehouse has room-sized bins containing upward of a thousand tons of the most popular building blocks, things like structural fibres. They're made at plants on the West Coast and shipped here by subway for ready use.

Other kinds are made on site using the special-purpose assemblers. Carl's main room has several cabinet-sized boxes hooked up to the plumbing, each taking in raw materials, running them through this sort of specialised molecular machinery, and pumping out a milky syrup of product. One syrup contains motors, another one contains computers, and another is full of microscopic plug-in light sources. All go into tanks for later use. Now they're being used. The mix for the Red Cross tent job is mostly structural fibre stronger than the old bulletproof-vest materials. Other building blocks also go in, including motors, computers, and dozens of little struts, angle brackets, and doohickies. The mix would look like someone had stirred together the parts from a dozen toy sets, if the parts were big enough to see. In fact, though, the largest parts would be no more than blurry dots, if you saw one under a normal optical microscope.

The mix also contains block-assemblers, floating free like everything else. These machines are big, about like an office building in our simulation view with the standard settings. Each has several jointed arms, a computer, and several plugs and sockets. These do the actual construction work.

To begin the build, pumps pour the mix into a manufacturing pond. The constant tumbling motions of microscopic things in liquids would be too disorganised for building anything so large as a tent, so the block-assemblers start grabbing their neighbours. Within moments, they have linked up to form a framework spread through the liquid. Now that they are plugged together, they divide up jobs, and get to work. Instructions pour in from Carl's desktop computer.

The block-assemblers use sticky grippers to pull specific kinds of building blocks out of the liquid. They use their arms to plug them together. For a permanent job, they would be using blocks that bond together chemically and permanently. For these temporary tents, though, the Red Cross design uses a set of standard blocks that are put together with amazingly ordinary fasteners: these blocks have snaps, plugs, and screws, though of course the parts are atomically perfect and the threads on the screws are single helical rows of atoms. The resulting joints weaken the tent's structure somewhat, but who cares? The basic materials are almost a hundred times stronger than steel, so there is strength to waste if it makes manufacturing more convenient.

Fibre segments snap together to make fabrics. Some segments contain motors and computers, linked by fibres that contain power and data cables. Struts snap together with more motors and computers to make the tent's main structures. Special surfaces are made of special building blocks. From the human perspective, each tent is a lightweight structure that contains most of the conveniences and comforts of an apartment: cooking facilities, a bathroom, beds, windows, air conditioning, specially modified to meet the environmental demands of the quake-stricken country. From a builder's perspective, especially from a nanomachine's point of view, the tent is just structure slapped together from a few hundred kinds of prefab parts.

In a matter of seconds, each block-assembler has put together a few thousand parts, and its section of the tent is done. In fact, the whole thing is done: many trillions of hands make light work. A crane swings out over the pond and starts plucking out tent packages as fresh mix flows in.

Maria's concern has drawn her back to the plant to see how the build is going. "It's coming along," Carl reassures her. "Look, the first batch of tents is out." In the warehouse, the first pallet is already stacked with five layers of dove-gray "suitcases": tents dried and packed for transport. Carl grabs a tent by the handle and lugs it out the door.

He pushes a tab on the corner labelled "Open," and it takes over a minute to unfold to a structure a half-dozen paces on a side. The tent is big, and light enough to blow away if it didn't cling to the ground so tightly. Maria and Carl tour the tent, testing the appliances, checking the construction of furniture: everything is extremely lightweight compared to the bulk-manufactured goods of the 1990s, tough but almost hollow.

Like the other structures, the walls and floors are full of tiny motors and struts controlled by simple computers like the ones used in twentieth-century cars, televisions, and pinball machines. They can unfold and refold. They can also flex to produce sound like a high-quality speaker, or to absorb sound to silence outdoors racket. The whole three-room setup is small and efficient, looking like a cross between a boat cabin and a Japanese business hotel room. Outside, though, it is little more than a box. Maria shakes her head, knowing full well what architects can do these days when they try to make a building really fit its site. Oh well, she thinks, These won't be used for long.

"Well, that looks pretty good to me," says Carl with satisfaction. "And I think we'll be finished in another hour." Maria is relieved. "I'm glad you had those pools freed up so fast." By three o'clock, they've shipped three thousand emergency shelters, sending them by subway. Within half an hour, tents are being set up at the disaster site.

Cheap Manufacturing

In the Desert Rose Industries scenario, manufacturing has become cheap, fast, clean, and efficient. Using fast, precise machines to handle matter in molecular pieces makes it easy for nanotechnology to be fast, clean, and efficient. But for it to be cheap, the manufacturing equipment has to be cheap.

The Desert Rose scenario shows how this can work. Molecular-manufacturing equipment can be used to make all the parts needed to build more molecular manufacturing equipment. It can even build the machines needed to put the parts together. This resembles an idea developed by NASA for a self-expanding manufacturing complex on the Moon, but made faster and simpler using molecular machines and parts.

General Assemblers

From the molecular manipulators and primitive assemblers described in the last chapter, the most likely path to nanotechnology leads to assemblers with more and more general capabilities. Still, efficiency favours special-purpose machines, and the Desert Rose scenario didn't make much use of general assemblers. Why bother making general-purpose assemblers in the first place?

To see the answer, turn the question around and ask, Why not build such a tool? There is nothing outstandingly difficult about a general assembler, as molecular machinery goes. It will just be a device with good, flexible positional control and a system to feed it a variety of molecular tools. This is a useful, basic capability. General-purpose assemblers could always be replaced by a lot of specialised devices, but to build those specialised devices in the first place, it makes sense to come up with a more flexible, general-purpose system that can just be reprogrammed. So, general purpose machines are likely to find use in making short production runs of more specialised devices. Ralph Merkle, a computers and security expert at Xerox Palo Alto Research Centre, sees this as paralleling the way manufacturing works today: "General purpose devices could do many tasks, but they'll do them inefficiently. For any given task, there will be one or a few best ways of doing it, and one or a few special purpose devices that are finely tuned to do that one task.

Nails aren't made by a general-purpose machine shop, they're made by nail-making machines. Making nails with a general-purpose machine shop would be more expensive, more difficult, and more time-consuming. Likewise, in the future we won't see a proliferation of general-purpose self-replicating systems, we'll see specialisation for almost every task."

Capabilities and Possibilities

We've surveyed a lot of devices: assemblers of various flavours, nanocomputers, disassemblers, replicators, and others. What's important about these is not the exact distinctions between them, but the capabilities that they will give and the effects they will have on human lives. Again, we are suspending discussion of potential misapplications until later. If we tease apart the implications of what we've seen in the Desert Rose scenario, we can analyze some of the key impacts of molecular manufacturing in industry, science, and medicine.

Technology and Industry

At its base, nanotechnology is about molecular manufacturing, and manufacturing is the basis of much of today's industry. This is why Desert Rose made a good starting point for describing the possibilities of a nanotechnological world. From an industrial perspective, it makes sense to think of nanotechnology in terms of products and production.

***New Products*:** Today, we handle matter crudely, but nanotechnology will bring thorough control of the structure of matter, the ability to build objects to atom-by-atom specifications. This means being able to make almost anything. By comparison, even today's range of products will feel very limited. Nanotechnology will make possible a huge range of new products, a range we can't envision today. Still, to get a feel for what is possible, we can look at some easily imagined applications.

***Reliable Products*:** Today, products often fail, but for failures to occur-for a wing to fall off an airplane, or a bearing to wear out-a lot of atoms have to be out of place. In the future, we can do better. There are two basic reasons for this: better materials and better quality control, both achieved by molecular manufacturing. By using materials tens of times stronger than steel, as Desert Rose did, it will be easy to make things that are very strong, with a huge safety margin. By building things with atom-by-atom control, flaws can be made very rare and extremely small-nonexistent, by present standards.

With nanotechnology, we can design in big safety margins and then manufacture the design with near-perfection. The result will be products that are tough and reliable. (There will still be room for bad designs, and for people who wish to take risks in machines that balance on the edge of disaster.)

Intelligent Products: Today, we make most things from big chunks of metal, wood, plastic, and the like, or from tangles of fibres. Objects made with molecular manufacturing can contain trillions of microscopic motors and computers, forming parts that work together to do something useful.

A climber's rope can be made of fibres that slide around and reweave to eliminate frayed spots. Tents can be made of parts that slide and lock to turn a package into a building. Walls and furniture can be made to repair themselves, instead of passively deteriorating.

On a mundane level, this sort of flexibility will increase reliability and durability. Beyond this, it will make possible new products with abilities we never imagined we needed so badly. And beyond even this, it will open new possibilities for art.

Inexpensive Production: Today, production requires a lot of labour, either for making things or for building and maintaining machines that make things. Labour is expensive, and expensive machines make automation expensive, too. In the Desert Rose scenario, we got a glimpse of how molecular manufacturing can make production far less expensive than it is today. This is perhaps the most surprising conclusion about nanotechnology, so we'll take a closer look at it in the next chapter.

Clean Production: Today, our manufacturing processes handle matter sloppily, producing pollution. One step puts stuff where it shouldn't be; the next washes it off the product and into the water supply. Our transportation system worsens the problem as unreliable trucks and tankers spill noxious chemicals over the land and sea. Everything is expensive, so companies skimp on even the half-effective pollution controls that we know how to build.

Nanotechnology will mean greater control of matter, making it easy to avoid pollution. This means that a little public pressure will go a long way toward a cleaner environment. Likewise, it will make it easy to increase efficiency and reduce resource requirements.

Products, like the Red Cross tents at Desert Rose, can be made of snap-together, easily recyclable parts. Sophisticated products could even be made from biodegradable materials. Nanotechnology will make it easy to attack the causes of pollution at their technological root.

Nanotechnology will have great applications in the field of industry, much as transistors had great applications in the field of vacuum tube electronics, and democracy had great applications in the field of monarchy. It will not so much advance twentieth-century industry as replace it-not all at once, but during a thin slice of historical time.

Science

Chemistry: Today, chemists work with huge number of molecules and study them using clever, indirect techniques. Making a new molecule can be a major project, and studying it can be another. Molecular manufacturing will help chemists make what they want to study, and it will help them make the tools they need to study it. Nanoinstruments will be used to prod, measure, and modify molecules in a host of ways, studying their structures, behaviours, and interactions.

Materials: Today, materials scientists make new superconductors, semiconductors, and structural materials by mixing and crushing and baking and freezing, and so forth. They dream of far more structures than they can make, and they stumble across more things than they plan. With molecular manufacturing, materials science can be much more systematic and thorough. New ideas can be tested because new materials can be built according to plan (rather than playing around, groping for a recipe).

This need not rule out unexpected discoveries, since experiments-even blind searches-will go much faster. A few tons of raw materials would be enough to make a billion samples, each a cubic micron in size. In all of history so far, materials scientists have never tested so many materials. With nanoinstruments and nanocomputers, they could. One laboratory could then do more than all of today's materials scientists put together.

Biology: Today, biologists use a host of molecular devices borrowed from biology to study biology. Many of these can be viewed as molecular machines. Nanotechnology will greatly advance biology by providing better molecular devices, better nanoinstruments. Some cells have already been mapped in amazing molecular detail, but biology still has far to go. With nanoinstruments (including molecule-by-molecule disassemblers), biologists will at last be able to map cells completely and study their interactions in detail. It will become easy not only to find molecules in cells, but to learn what they do. This will help in understanding disease and the molecular requirements for health, enormously advancing medicine.

Computation: Today, computers range from a million to a billion times faster than an old desktop adding machine, and the results have been revolutionary for science. Every year, more questions can be answered by calculations based on known principles of physics. The advent of nanocomputers-even slow, miserable, mechanical nanocomputers-will give us practical machines with a trillion times the power of today's computers (essentially by letting us package a trillion computers in a small space, without gobbling too much money or energy.) The consequences will again be revolutionary.

Physics: The known principles of physics are adequate for understanding molecules, materials, and cells, but not for understanding phenomena on a scale that would still be submicroscopic if atoms were the size of marbles. Nanotechnology can't help here directly, but it can provide manufacturing facilities that will make huge particle accelerators economical, where today they strain national budgets.

More generally, nanotechnology will help science wherever precision and fine details are important. Science frequently proceeds by trying small variations in almost identical experiments, comparing the results. This will be easier when molecular manufacturing can make two objects that are identical, molecule by molecule. In some areas, today's techniques are not only crude, but destructive.

Archaeological sites are unique records of the human past, but today's techniques throw away most information during the dig, by accident. Future archaeologists, able to sift soil not speck by speck but molecule by molecule, will be grateful indeed to those archaeologists who today leave some ground undisturbed.

Medicine

Of all the areas where the ability to manufacture new tools is important, medicine is perhaps the greatest. The human body is intricate, and that intricacy extends beyond the range of human vision, beyond microscopic imaging, down to the molecular scale. "Molecular medicine" is an increasingly popular term today, but medicine today has only the simplest of molecular tools.

As biology uses nanoinstruments to learn about disease and health, we will learn the physical requirements for restoring and maintaining health. And with this knowledge will come the tools needed to satisfy those requirements-tools ranging from improved pharmaceuticals to devices able to repair cells and tissues through molecular surgery.

Advanced medicine will be among the most complex and difficult applications of nanotechnology. It will require great knowledge, but nanoinstruments will help gather this knowledge. It will pose great engineering challenges, but computers of trillionfold greater power will help meet those challenges. It will solve medical problems on which we spend billions of dollars today, in hopes of modest improvements.

Today, modern medicine often means an expensive way to prolong misery. Will nanomedicine be more of the same? Any reader over the age of, say, thirty knows how things start to go wrong: an ache here, a wrinkle there, the loss of an ability. Over the decades, the physical quality of life declines faster and faster-thc limits of what the body can do become stricter-until the limits are those of a hospital bed. The healing abilities we have when young seem to fade away. Modern medical practice expends the bulk of its effort on such things as intensive care units, dragging out the last few years of life without restoring health.

Truly advanced medicine will be able to restore and supplement the youthful ability to heal. Its cost will depend on the cost of producing things more intricate than any we have seen before, the cost of producing computers, sensors, and the like by the trillions. To understand the prospects for medicine, like those for science and industry, we need to take a closer look at the cost of molecular manufacturing.

Applicability of Nanotechnology

Over the past few decades, the fields of science and engineering have been seeking to develop new and improved upon types of energy technologies that have the capability of improving life all over the world. In order to make the next leap forward from the current generation of technology, scientists and engineers have been developing a new field of science called Nanotechnology. Nanotechnology is simply any technology that contains components smaller than 100 nanometres. For scale, a single virus particle is about 100 nanometres in width.

An important subfield of Nanotechnology related to energy is Nanofabrication. Nanofabrication is the process of designing and creating devices on the nanoscale. Creating devices smaller than 100 nanometres opens many doors for the development of new ways to capture, store, and transfer energy.

The inherent level of control that Nanofabrication could give scientists and engineers would be critical in providing the capability of solving many of the problems that the world is facing today related to the current generation of energy technologies.

People in the fields of science and engineering have already begun developing ways of utilising Nanotechnology for the development of consumer products. Benefits already observed from the design of these products are an increased efficiency of lighting and heating, increased electrical storage capacity, and a decrease in the amount of pollution from the use of energy. Benefits such as these make the investment of capital in the research and development of Nanotechnology a top priority.

Consumer Products

Recently, previously established and entirely new companies such as BetaBatt, Inc. and Oxane Materials are focusing on Nanomaterials as a way to develop and improve upon older methods for the capture, transfer, and storage of energy for the development of consumer products.

ConsERV, a product developed by the Dais Analytic Corporation, uses nanoscale polymer membranes to increase the efficiency of heating and cooling systems and has already proven to be a lucrative design. The polymer membrane was specifically configured for this application by selectively engineering the size of the pores in the membrane to prevent air from passing, while allowing moisture to pass through the membrane.

Polymer membranes can be designed to selectively allow particles of one size and shape to pass through while preventing others of different dimensions. This makes for a powerful tool that can be used in consumer products from biological weapons protection to industrial chemical separations.

A New York based company called Applied NanoWorks, Inc. has been developing a consumer product that utilises LED technology to generate light. Light Emitting Diodes or LEDs, use only about 10% of the energy that a typical incandescent or fluorescent light bulb use and typically lasts much longer, which makes them a viable alternative to traditional light bulbs. While LEDs have been around for decades, this company and others like it have been developing a special variant of LED called the white LED.

White LEDs consist of semi-conducting organic layers that are only about 100 nanometres in distance from each other and are placed between two electrodes, which create an anode, and a cathode. When voltage is applied to the system, light is generated when electricity passes through the two organic layers. This is called electroluminescence. The semiconductor properties of the organic layers are what allow for the minimal amount of energy necessary to generate light. In traditional light bulbs, a metal filament is used to generate light when electricity is run through the filament. Using metal generates a great deal of heat and therefore lowers efficiency.

Research for longer lasting batteries has been an ongoing process for years. Researchers have now begun to utilise Nanotechnology for battery technology. mPhase Technologies in conglomeration with Rutgers University and Bell Laboratories have utilised nanomaterials to alter the Wetting Behaviour of the surface where the liquid in the battery lies to spread the liquid droplets over a greater area on the surface and therefore have greater control over the movement of the droplets. This gives more control to the designer of the battery. This control prevents reactions in the battery by separating the electrolytic liquid from the anode and the cathode when the battery is not in use and joining them when the battery is in need of use.

Economic Benefits

The relatively recent shift toward using Nanotechnology with respect to the capture, transfer, and storage of energy has and will continue to have many positive economic impacts on society. The control of materials that

Nanotechnology offers to scientists and engineers of consumer products is one of the most important aspects of Nanotechnology. This allows for an improved efficiency of products across the board.

A major issue with current energy generation is the loss of efficiency from the generation of heat as a by-product of the process. A common example of this is the heat generated by the internal combustion engine.

The internal combustion engine losses about 36% of the energy from gasoline as heat and an improvement of this alone could have a significant economic impact. However, improving the internal combustion engine in this respect has proven to be extremely difficult without sacrificing performance. Improving the efficiency of fuel cells through the use of Nanotechnology appears to be more plausible by using molecularly tailored catalysts, polymer membranes, and improved fuel storage.

In order for a fuel cell to operate, particularly of the hydrogen variant, a noble-metal catalyst (usually platinum, which is very expensive) is needed to separate the electrons from the protons of the hydrogen atoms. However, catalysts of this type are extremely sensitive to carbon monoxide reactions. In order to combat this, alcohols or hydrocarbons compounds are used to lower the carbon monoxide concentration in the system. This adds an additional cost to the device. Using nanotechnology, catalysts can be designed through nanofabrication that are much more resistant to carbon monoxide reactions, which improves the efficiency of the process and may be designed with cheaper materials to additionally lower costs.

Fuel cells that are currently designed for transportation need rapid start-up periods for the practicality of consumer use. This process puts a lot of strain on the traditional polymer electrolyte membranes, which decreases the life of the membrane requiring frequent replacement. Using Nanotechnology, engineers have the ability to create a much more durable polymer membrane, which addresses this problem. Nanoscale polymer membranes are also much more efficient in ionic conductivity. This improves the efficiency of the system and decreases the time between replacements, which lowers costs.

Another problem with contemporary fuel cells is the storage of the fuel. In the case of hydrogen fuel cells, storing the hydrogen in gaseous rather than liquid form improves the efficiency by 5%.

However, the materials that we currently have available to us significantly limit fuel storage due to low stress tolerance and costs. Scientists have come up with an answer to this by using a nanoporous styrene material (which is a relatively inexpensive material) that when super-cooled to around-196oC, naturally holds on to hydrogen atoms and when heated again releases the hydrogen for use.

Capacitors

For decades, scientists and engineers have been attempting to make computers smaller and more efficient. A crucial component of computers are capacitors. A capacitor is a device that is made of a pair of electrodes separated by an insulator that each stores an opposite charge.

A capacitor stores a charge when it is removed from the circuit that it is connected to; the charge is released when it is replaced back into the circuit. Capacitors have an advantage over batteries in that they release their charge much more quickly than a battery.

Traditional or foil capacitors are composed of thin metal conducting plates separated by an electrical insulator, which are then stacked or rolled and placed in a casing. The problem with a traditional capacitor such as this is that they limit how small an engineer can design a computer. Scientists and engineers have since turned to Nanotechnology for a solution to the problem.

Using nanotechnology, researchers developed what they call "ultracapacitors." An ultracapacitor is a general term that describes a capacitor that contains nanocomponents. Ultracapacitors are being researched heavily because of their high density interior, compact size, reliability, and high capacitance.

This decrease in size makes it increasingly possible to develop much smaller circuits and computers. Ultracapacitors also have the capability to supplement batteries in hybrid vehicles by providing a large amount of energy during peak acceleration and allowing the battery to supply energy over longer periods of time, such as during a constant driving speed. This could decrease the size and weight of the large batteries needed in hybrid vehicles as well as take additional stress off the battery. However, as of now, the combination of ultracapacitors and a battery is not cost affective due to the need of additional DC/DC electronics to coordinate the two.

Nanoporous carbon aerogel is one type of material that is being utilised for the design of ultracapacitors. These aerogels have a very large interior surface area and can have its properties altered by changing the pore diameter and distribution along with adding nanosized alkali metals to alter its conductivity.

Carbon nanotubes are another possible material for use in an ultracapacitor. Carbon nanotubes are created by vaporizing carbon and allowing it to condense on a surface. When the carbon condenses, it forms a nanosized tube composed of carbon atoms. This tube has a high surface area, which increases the amount of charge that can be stored. The low reliability and high cost of using carbon nanotubes for ultracapacitors is currently an issue of research.

Capacitance Theory

Understanding the concept of capacitance can be helpful in understanding why Nanotechnology is such a powerful tool for the design of higher energy storing capacitors. A capacitor's capacitance (C) or amount of energy stored is equal to the amount of charge (Q) stored on each plate divided by the voltage (V) between the plates.

Another representation of capacitance is that capacitance (C) is approximately equal to the permittivity (e) of the dielectric times the area (A) of the plates divided by the distance (d) between them. Therefore, capacitance is proportional to the surface area of the conducting plate and inversely proportional to the distance between the plates.

Using carbon nanotubes as an example, a property of carbon nanotubes is that they have a very high surface area to store a charge. Using the above proportionality that capacitance (C) is proportional to the surface area (A) of the conducting plate; it becomes obvious that using nanoscaled materials with high surface area would be great for increasing capacitance.

The other proportionality described above is that capacitance (C) is inversely proportional to the distance (d) between the plates. Using nanoscaled plates such as carbon nanotubes with nanofabrication techniques, gives the capability of decreasing the space between plates which again increases capacitance.

Centre on Nanotechnology and Society

The Centre on Nanotechnology and Society (Nano & Society) is an affiliate of Illinois Institute of Technology (IIT) and is housed at IIT's Chicago-Kent College of Law. Nano & Society is an affiliate of the Institute on Biotechnology and the Human Future, also based at Chicago-Kent College of Law.

Purpose

The Centre on Nanotechnology and Society (Nano & Society) was created to catalyze informed interdisciplinary research, education, and dialogue on the ethical, legal, economic, policy, and broader societal implications of nanoscale science and technology. As an affiliate of the Institute on Biotechnology and the Human Future, Nano & Society approaches nanotechnology with a special focus on the human condition.

Nano & Society brings the foremost scholars and researchers in nano science, law, ethics, medicine, and the social sciences together with leaders in business and industry. Nano & Society also employs a number of vehicles in its efforts to focus a national conversation on nanotechnology and human dignity, including its website, the Chicago Nano Forum, national conferences, and NELSI Global (a public posicy document archive).

Online Offerings

Nano & Society conducts research on NELSI—Nanotechnology's ethical, legal, and social implications. The Centre features various initiatvies: NELSI Global is a web-based global public policy document archive focused on nanotechnology's ethical, legal and societal implications (NELSI). This initiative is accessible through Nano & Society's website. NELSI Global serves as a unified clearinghouse, enabling users to identify and access key public policy documents addressing the NELSI agenda. It includes international and U.S.-specific regulations, legislation, case law, congressional testimony, and key governmental reports.

A Website that Addresses: law, policy, and regulation; environment, health and safety; business; applications; and Publications

Publications, Conferences, and Events

The Centre on Nanotechnology and Society uses multiple communications vehicles to educate and dissemi-nate information on NELSI. These mechanisms include:

Nano and Society: The Centre on Nanotechnology and Society publishes a monthly e-newsletter, entitled Nano & Society. Which features opinion pieces from guest writers on nanotechnology's potential and potential impact, as well as public attitudes toward and education about nanotechnology; and relevant news updates on nanotechnology.

Nanologues: The Centre on Nanotechnology and Society publishes a print series of booklets on NELSI entitled Nanologues.

Conferences and Events: Nano & Society hosts various national events to advance the dialogue on nanotechnology and the human condition, including:

Chicago Nano Forum: In conjunction with its partners in the educational, business, environmental, legal, and other arenas, Nano & Society has developed the Chicago Nano Forum. The Forum encourages public dialogue among nano experts in science, business, social sciences, ethics, and law. This series of events is webcast on the Centre's website.

1st Annual Conference on Nanopolicy and the Human Future: This Nano & Society conference provided members of Congress and their staff, researchers, scientists, and others interested in and involved with the nano sector with the latest nanotechnology developments in the ethical, legal and social arena.

Centre for Responsible Nanotechnology

The Centre for Responsible Nanotechnology (CRN), founded in December, 2002, is a non-profit research and advocacy organisation with a focus on molecular manufacturing and its possible effects, both positive and negative.

CRN provides information to journalists, business leaders, policymakers and the general public about the environmental, humanitarian, economic, military, political, social, medical, and ethical implications of advanced nanotechnology.

CRN is an affiliate of World Care, an international non-profit 501(c)(3) organisation. As of 2006, CRN's Director of Research is Chris Phoenix, and its Executive Director is Mike Treder. In August 2005, a task force of more than sixty international experts from various fields was organised by CRN to develop comprehensive recommendations for the safe and responsible use of nanotechnology.

A series of eleven essays written by members of this CRN Task Force was published in March 2006 by the journal Nanotechnology Perceptions. The Centre has sometimes been criticised for concentrating too heavily on the dangers posed by nanotechnology.

National Institute for Nanotechnology

The National Institute for Nanotechnology (NINT) is a Canadian research institution located on the University of Alberta main campus, in Edmonton, Alberta, Canada. Its primary purpose is nanotechnological research.

The institute was established in 2001 as a partnership between the National Research Council of Canada and the University of Alberta, and is funded by the Government of Canada, the Government of Alberta and the university. In June 2006, the institute moved into their present 20,000 square metre facility, designed to be one of the world's largest buildings for nanotechnological research. There are at most two or three other facilities worldwide matching the new building in scale and capacity. According to NINT's research plan, the institute's research focus is in the following areas:

- Synthesis and characterization of nanocrystals and nanowires
- Synthesis of supramolecular-based nanomaterials
- Fabrication and characterization of molecular-scale devices and nanosensors
- Development of nano-scaled materials for catalysis and directed chemical reactions at semiconductor surfaces
- Development of nano-electronic and nano-fluidic systems to interface devices to the outside world
- Theory, modelling, and simulation of nanosystems on multiple length scales
- Development of quantitative imaging and characterization techniques that support nanotechnology research

The work is focused around four interdisciplinary research groups:

- Molecular scale development
- Supramolecular nanoscale assembly
- Materials and interfacial chemistry
- Theory and modelling

Complex Nanomaterials

To understand complex nanomaterials or nanomaterials having complex composition and their utility, a short narrative on nanomaterials is provided. Nanomaterials are materials possessing one or more dimensional features having a length on the order of a billionth of a metre to less than 100 billionths of a metre. They are important because (1) they exhibit unique properties which are derived from the size of these features and (2) we have fairly recently learned how to manipulate matter on these dimensions so as to understand and exploit their unique properties and their relationship to size. The properties of matter are largely a result of their composition, and the conditions under which they were produced.

The properties of most materials are largely a result of variations in composition, temperature and pressure and the rates at which these independent parameters are changed during their synthesis or production. As nanotechnologists, we look at "size" as an independent degree of freedom which can be manipulated independent of composition, temperature and pressure to yield materials that possess new properties not exhibited by their conventional counterparts. When materials possesses size features that are on the order of a few billionths of a metre, those materials often exhibit new properties not found in their ordinary material counterparts.

The NNI website which puts these ideas in a simple phrase, "At the nanoscale, the physical, chemical, and biological properties of materials differ in fundamental and valuable ways from the properties of individual atoms and molecules or bulk matter." This phrase can be further simplified to read: At the nanoscale, physical, chemical, and biological properties differ from the properties of individual atoms and molecules or bulk matter.

The nuances of how a material is produced often affects the molecular structure and the possible structure so produced give rise to variations in the materials bulk properties. An example of this is the case hardening of steel. A hard skin is created over a softer core by heating the steel to a high temperature and then rapidly cooling (quenching) it. The resulting composite structure, hard skin/softer core that develops is a function of the rate at which the temperature changed on the surface as compared to that in the core. (This will be discussed in greater detail below.)

To understand nanomaterials, normal materials (non-nanomaterials) are described to provide context. Materials are composed of atoms, molecules, ions, compositions thereof and assemblies of other materials. Solid materials can be broadly classified into crystalline and amorphous.

Crystalline materials come in two main varieties, single crystalline and polycrystalline. A crystalline material is composed of an orderly array of atoms, molecules or ions. Crystalline materials generally have short and long range order. That means that the manner in which atoms are arranged at any one location within a crystal is identical to the arrangement of at any other location. An example of a crystalline material is ordinary table salt.

Each grain of salt is a single crystal. A diamond is also a single crystal. The semiconductor or chip inside a computer's microprocessor is a single crystal of silicon. An example of a polycrystalline material is a ceramic dish or coffee cup. If the broken edge of such item is examined with a magnifying glass one can easily see the individual crystals, referred to as crystallites or grains.

When a metal breaks, the grains or crystallites of which it is composed can be observed. In a figure coming soon is a metallograph of a polished metal surface. The grains are easily observed. The surface between grains is called the "grain boundary," and they appear as the line between grains in Figure coming soon. All crystalline materials exhibit Bragg scattering of x-rays. Amorphous materials are non crystalline. They have short range order but lack long range order. Polymers such as polyethylene, polyester and polypropylene are amorphous. Glass is an amorphous solid, which has short range order (always a Si next to an O, etc.) but lacks long range order. Amorphous materials are usually not composed of grains.

Nanomaterials include materials where the size of the particles, crystallites or grains of which the material is composed are on the order of nanometres. As a result of this properties emerge that are not characteristic of their ordinary counterparts having the same composition.

These properties (examples given below) some from the fact that within each grain there is a relatively small number of atoms, molecules or ions as compared to a normal material. As the particle or grain size is in the nanometre range, many of the atoms or each grain or within the material as a whole reside at the particle surface, or grain boundary. The number of particle that reside at the surface or grain boundary is proportional to 1/r where r is the average radius of the particle size or grain.

That means that, as the particle size becomes smaller the number of atoms, molecules or ions that exist at the surface or grain boundary increases. Accordingly, the properties of a large collection of nanomaterials are dominated by the properties of the grain boundary or surface. As an example of how the size affects properties consider a material that exhibits ordered electronic spin on the surface but disordered electronic spin in the interior of the particle.

As the particle is reduced in size, the number of atoms on the surface increases as compared to those in the interior and so the number of ordered electronic spins increases as the particle size decreases. A material completely composed of nanoparticles or nano-crystallites of this type would exhibit properties of derived from the ordered spins, e.g. magnetism, while a normal material having the same composition would not exhibit such a property.

Nanomaterials can exist in several forms. The simplest are nanopowders. Nanopowders can be consolidated so as to form a polycrystalline nanomaterials. They can also be combined with other materials to form composite materials.

COMPARATIVE ANALYSIS

Complex materials are analogous to steel or other metal alloys because of their enhanced performance characteristics, while simple materials are more analogous to pure iron, aluminum or copper. An alloy is a mixture of several elements such as in the case of stainless steel: iron, nickel and chromium, each component of which imparts important characteristics to the performance of the alloy.

A pure metal is predominantly composed of a single element, for example iron, as in an iron skillet or heating radiator found in homes built in the first half of the 20th century in this country. The first train rails were made out of iron, while all modern rails use steel alloys. Why? Because modern rails last longer, carry many times more weight, are less susceptible to corrosion, and will not become brittle and crack with changing temperatures. Virtually all modern metals in ordinary use are made out of alloys that improve performance, strength and longevity.

Early successes in the nanomaterials industry (use of nano particles in sun screen and ultra fine polishing of semiconductor wafers) represent simple material applications and in these applications only rudimentary aspects of nanotechnology are exploited. Based on current performance limitations, future developments in fuel cells, solar technology, electronics, optics, biotechnology, cosmetics and advanced structural materials involve solutions requiring complex nanomaterials.

Nanotoxicology

Nanotoxicology is the study of the toxicity of nanomaterials. Because of the small size and large surface area of nanomaterials, these materials have unique properties compared with their larger counterparts. The nanomaterials, even when they are made of inert elements like gold, become very active at a nanometre range. Nanotoxicological studies are intended to determine whether and to what extent these may pose a threat to the environment and to human beings.

Research on ultrafine particles has laid the foundation for the emerging field of nanotoxicology, with the goal of studying the biokinetics of engineered nanomaterials and their potential for causing adverse effects. Most reports find that ultrafine particles are more toxic than equivalent larger-sized particles of a given material at similar doses per gram of body weight.

Assessing the safety of engineered nanoparticles is a highly complex matter that goes beyond traditional toxicology. Engineered nanoparticles are not a uniform group of substances. The problem arises from the fact that particle size alone (which determines the surface area of a given mass of a substance) is not the only factor that determines the toxicological impact of a material and that makes nanoparticles of a given substance behave differently from the bulk form.

While the release and production of nanoparticles during industrial and combustion processes and activities is mostly unintentional, the emergence of nanotechnological production processes introduces the intentional and controlled manufacture of nanoparticles. The latter can be further differentiated as either bulk nanoparticles in industry, e.g., carbon black or titanium dioxide, or so-called engineered nanoparticles, e.g., carbon nanotubes.

Given the prospects for nanotechnology, and the fact that products containing engineered nanoparticles have already been introduced to the marketplace, the increasing flow of new products will bring about the massive production of engineered nanoparticles.

Despite all the scientific knowledge gained in the toxicology of particulate matter, because of the many variables involved, scientists still cannot accurately predict how nanomaterials will affect living organisms. What is clear, though, is that the biologic activity and biokinetics of nanoparticles are different from larger particles, and that they depend on many parameters. These parameters can modify cellular uptake, protein binding, translocation from portal of entry to target site, and the possibility of tissue injury.

Important from a toxicological point of view are the physiochemical properties that come with size and lead to certain biological reactivity. Size, together with differences in shape, surface structure, chemical composition (purity, crystallinity, electrical properties, etc.), solubility and biopersistence make for a large number of variables that need to be considered when assessing nanoparticles. On top of that, powders or liquids containing nanoparticles are almost never monodisperse but contain a range of particle sizes. This complicates experimental analysis as larger nanoparticles might have different properties than smaller ones. Also, nanoparticles show a tendency to aggregate and such aggregates often behave differently from individual nanoparticles.

Most, if not all, toxicological studies on nanoparticles rely on current methods, practices and terminology as gained and applied in the analysis of micro-and ultrafine particles and mineral fibres. Together with recent studies on nanoparticles, this provides an initial basis for evaluating the primary issues in a risk assessment framework for nanomaterials.

Given the many parameters involved, nanotoxicology requires an interdisciplinary team approach, even more so than classical toxicology, in order to arrive at an appropriate risk assessment.

As a still-maturing science, nanotoxicology will expand the boundaries of traditional toxicology from a testing and auxiliary science to a new discipline where toxicological knowledge of nanomaterials can be put to constructive use in therapeutics as well as the development of new and better biocompatible materials.

Nanotribology

Nanotribology is a branch of tribology which studies friction phenomenon at the nanometre scale. The distinction between nanotribology and tribology is primarily due to the involvement of atomic forces in the determination of the final behaviour of the system.

Gears, bearings, and liquid lubricants can reduce friction in the macroscopic world, but the origins of friction for small devices such as micro-or nanoelectromechanical systems (NEMS) require other solutions. Despite the unprecedented accuracy by which these devices are nowadays designed and fabricated, their enormous surface-volume ratio leads to severe friction and wear issues, which dramatically reduce their applicability and lifetime. Traditional liquid lubricants become too viscous when confined in layers of molecular thickness. This situation has led to a number of proposals for ways to reduce friction on the nanoscale, such as superlubricity and thermolubricity.

Nanophotonics

Nanophotonics is the study of the behaviour of light on the nanometre scale. The ability to fabricate devices in nanoscale that has been developed recently provided the catalyst for this area of study.

Nanophotonics scientists are experimenting with different ways to generate and manipulate light using ultrasmall, engineered structures at the nanoscale. Tiny gold particles called "nanostars" are being studied to ascertain how they interact with light. The study has the potential to revolutionise the telecommunications industry by providing low power, high speed, interference-free devices such as electrooptic and all-optical switches on a chip. There are a number of research groups working in this area in the U.S., the UK, Italy and Spain.

Mechanochemistry

Mechanochemistry, is the coupling of the mechanical and the chemical on a molecular scale includes mechanical breakage, polymer degradation under shear, cavitation related phenomena (e.g., sonochemistry and sonoluminescence), shockwave chemistry and physics, and even the burgeoning field of molecular machines.

A small part of mechanochemistry is sometimes also called "positional synthesis" or "positional assembly" is a technique for forming chemical bonds by direct computer control of the position of molecules. This is an example of a specific type of Mechanosynthesis.

As of 2004, the typical experimental arrangement is to attach a molecule to the tip of an atomic force microscope, and then use the microscope's precise positioning abilities to push the molecule on the tip into another on a substrate. Since the angles and distances can be precisely controlled, and the reaction occurs in a vacuum, novel chemical compounds and arrangements are possible. Much of the excitement regarding mechanochemistry regards its potential use in automated assembly of molecular-scale devices. Such techniques appear to have many applications in medicine, aviation, resource extraction, manufacturing and warfare.

Most theoretical explorations of such machines have focused on using Carbon, because of the many strong bonds it can form, the many types of chemistry these bonds permit, and utility of these bonds in medical and mechanical applications. Carbon forms diamond, for example, which if cheaply available, would be an excellent material for many machines. In practice, getting exactly one molecule to a known place on the microscope's tip is possible, but has proven difficult to automate. Since practical products require at least several hundred million atoms, this technique has not yet proven practical in forming a real product.

The goal of mechanoassembly research at this point focuses on overcoming these problems by calibration, and selection of appropriate synthesis reactions. The first product to be built by these means will probably be a specialised, very small (roughly 1,000 nanometres on a side) machine tool that can build copies of itself using mechanochemical means, under the control of an external computer. In the literature, such a tool is called an assembler. Once assemblers exist, geometric growth (copies making copies) could reduce the cost of assemblers rapidly. Control by an external computer should then permit large groups of assemblers to construct large, useful projects to atomic precisions.

One such project would combine molecular-level conveyor belts with permanently-mounted assemblers to produce a factory.

Mechanochemistry actually goes back more than 100 years. The coupling of the mechanical and the chemical on a molecular scale includes mechanical breakage, polymer degradation under shear, cavitation related phenomena (e.g., sonochemistry and sonoluminescence), shockwave chemistry and physics, and even the burgeoning field of molecular machines.

The technique of moving single atoms mechanically was proposed by Eric Drexler in his 1986 book The Engines of Creation. In 1988, researchers at IBM's Zürich Research Institute successfully spelled the letters "IBM" in Xenon atoms on a cryogenic copper surface, grossly validating the approach. Since then, a number of research projects have undertaken to use similar techniques to store computer data in a compact fashion. More recently the technique has been used to explore novel physical chemistries, sometimes using lasers to excite the tips to particular energy states, or examine the quantum chemistry of particular chemical bonds.

Nano-Optics

Nano-optics is the branch of optical engineering which deals with optics at deeply subwavelength length scales. Technologies in the realm of nano-optics include near-field scanning optical microscopy (NSOM), photoassisted scanning tunnelling microscopy, and surface plasmon optics. Traditional microscopy makes use of diffractive elements to focus light tightly in order to increase resolution.

But because of the diffraction limit (also known as the Rayleigh Criterion), propagating light may be focused to a spot with a minimum diameter of roughly half the wavelength of light. Thus, even with diffraction-limited confocal microscopy, the maximum resolution obtainable is on the order of a couple hundred nanometres. The scientific and industrial communities are becoming more interested in the characterization of materials and phenomena on the scale of a few nanometres, so alternative techniques must be utilised. Scanning Probe Microscopy (SPM) makes use of a "probe," (usually either a tiny aperture or super-sharp tip), which either locally excites a sample or transmits local information from a sample to be collected and analyzed.

Nanolanguage

NanoLanguage is a new way of thinking scientific computing, combining the strength of flexible object-oriented scripting interfaces (known from Mathematica and MatLab) with sophisticated high performance scientific computing algorithms. The goal is to enable scientists to efficiently extend, specialise and combine methods to calculate nanoscale properties of matter, including density functional theory, semi-empirical tight-binding, classical potentials, k.p and various quantum-chemical methods.

NanoLanguage allows for both low and high level detailed control of the computer simulations performed in the Atomistix ToolKit. At the high level, it offers a common interface for setting up complex atomic-scale simulations and analyzing the results. On the lower level side, it provides an interface to the low-level functionality in ATK.

NanoLanguage is built on top of Python, a powerful and well-established interpreted programming language, and thus it includes basic elements such as loops over simulation control parameters, plus support for efficient manipulations of e.g. numerical array data. It is therefore an ideal tool for automating series of simulations where geometric, material, or other parameters are to be optimized.

Using NanoLanguage allows scientists to express models of nature in a common language without the need to re-implement already implemented algorithms, and it allows for third-party development of new functionality on top of the ATK platform. Such functionality may consist of new atomic-scale modelling methodologies, tailored semi-empirical methods, or complex post-processing methods for calculating new quantities from the fundamental simulation results.

The Nanomanufacturing

Nanomanufacturing is used to describe either the production of nano scaled materials, which can be powders or fluids, or describe the manufacturing of parts "button up" from nanoscaled materials or "top down" in smallest steps for high precison by several technologis like laser ablation, etching and others.

The term is widely used, e.g. by the European Technology Platform MINAM and the U.S. National Nanotechnology Initiative (NNI). NNI refer to a sub-effort of nanotechnology—one of its five "priority areas."

There is also a Nanomanufacturing Program at the U.S. National Science Foundation. Nanomanufacturing should not be confused with molecular manufacturing which refers specifically to the manufacture of complex, nanoscale structures by means of nonbiological mechanosynthesis (and subsequent assembly).

The NNI has defined nanotechnology very broadly, to include a wide range of tiny structures including those created by large and imprecise tools.

Nanomanufacturing is not defined in the NNI's recent report, "Instrumentation and Metrology for Nanotechnology." (For contrast, another "priority area," nanofabrication, is defined as "the ability to fabricate, by directed or self-assembly methods, functional structures or devices at the atomic or molecular level." p. 67) Nanomanufacturing appears to be the near-term industrial-scale manufacture of nanotechnology-based objects, with emphasis on low cost and reliability.

Catalyst Based on Nanomaterial

Nanomaterial-based catalysts are usually heterogeneous catalysts broken up into nanoparticles in order to speed up the catalytic process. The extremely small size of the particles maximizes the surface area exposed to the reactant, allowing more reactions to occur at the same time, thus speeding up the process. Much research on nanomaterial-based catalysts have to do with maximizing the effectiveness of the catalyst coating in fuel cells. Platinum is currently the most common catalyst for this application, however, it is expensive and rare, so a lot of research has been going into maximizing the catalytic properties of other metals by shrinking them to nanoparticles in the hope that someday they will be an efficient and economic alternative to platinum.

Defence programs in many countries are now concentrating on nanotechnology research that will facilitate advances in such technology used to create secure but small messaging equipment, allow the development of smart weapons, improve stealth capabilities, aid in developing specialised sensors (including bio-inclusive sensors), help to create self-repairing military equipment, and improve the development and delivery mechanisms for medicines and vaccines. Nanotechnology builds on advances in microelectronics during the last decades of the twentieth century. The miniaturization of electrical components greatly increased the utility and portability of computers, imaging equipment, microphones, and other electronics. Indeed, the production and wide use of such commonplace devices such as personal computers and cell phones was absolutely dependent on advances in microtechnology.

Despite these fundamental advances there remain real physical constraints (e.g., microchip design limitations) to further miniaturization based upon conventional engineering principles. Nanotechnologies intend to revolutionize components and manufacturing techniques to overcome these fundamental limitations. In addition, there are classes of biosensors and feedback control devices that require nanotechnology because-despite advances in microtechnology-present components remain too large or slow.

Advances

Nanotechnology advances affect all branches of engineering and science that deal directly with device components ranging in size between 1/10,000,000 (one ten millionth of a millimetre) and 1/10,0000 millimetre. At these scales, even the most sophisticated microtechnology-based instrumentation is useless. Engineers anticipate that advances in nanotechnology will allow the direct manipulation of molecules in biological samples (e.g., proteins or nucleic acids) paving the way for the development of new materials that have a biological component or that can provide a biological interface.

In addition to new tools, nanotechnology programs advance practical understanding of quantum physics. The internalization of quantum concepts is a necessary component of nanotechnology research programs because the laws of classical physics (e.g., classical mechanics or generalized gas laws) do not always apply to the atomic and near-atomic level. Nanotechnology and quantum physics. Quantum theory and mechanics describe the relationship between energy and matter on the atomic and subatomic scale.

At the beginning of the twentieth century, German physicist Maxwell Planck (1858-1947) proposed that atoms absorb or emit electromagnetic radiation in bundles of energy termed quanta. This quantum concept seemed counter-intuitive to well-established Newtonian physics. Advancements associated with quantum mechanics (e.g., the uncertainty principle) also had profound implications with regard to the philosophical scientific arguments regarding the limitations of human knowledge.

Planck's quantum theory, which also asserted that the energy of light (a photon) was directly proportional to its frequency, proved a powerful concept that accounted for a wide range of physical phenomena. Planck's constant relates the energy of a photon with the frequency of light. Along with the constant for the speed of light, Planck's constant (h = 6.626 x 10-34 Joule-second) is a fundamental constant of nature.

Prior to Planck's work, electromagnetic radiation (light) was thought to travel in waves with an infinite number of available frequencies and wavelengths. Planck's work focused on attempting to explain the limited spectrum of light emitted by hot objects. Danish physicist Niels Bohr (1885-1962) studied Planck's quantum theory of radiation and worked in England with physicists J. J. Thomson (1856-1940), and Ernest Rutherford (1871-1937) to improve their classical models of the atom by incorporating quantum theory. During this time, Bohr developed his model of atomic structure.

According to the Bohr model, when an electron is excited by energy it jumps from its ground state to an excited state (i.e., a higher energy orbital). The excited atom can then emit energy only in certain (quantized) amounts as its electrons jump back to lower energy orbits located closer to the nucleus. This excess energy is emitted in quanta of electromagnetic radiation (photons of light) that have exactly the same energy as the difference in energy between the orbits jumped by the electron.

The electron quantum leaps between orbits proposed by the Bohr model accounted for Plank's observations that atoms emit or absorb electromagnetic radiation in quanta. Bohr's model also explained many important properties of the photoelectric effect described by Albert Einstein (1879-1955). Einstein assumed that light was transmitted as a stream of particles termed photons. By extending the well-known wave properties of light to include a treatment of light as a stream of photons, Einstein was able to explain the photoelectric effect. Photoelectric properties are key to regulation of many microtechnology and proposed nanotechnology level systems.

Quantum mechanics ultimately replaced electron "orbitals" of earlier atomic models with allowable values for angular momentum (angular velocity multiplied by mass) and depicted electron positions in terms of probability "clouds" and regions.

In the 1920s, the concept of quantization and its application to physical phenomena was further advanced by more mathematically complex models based on the work of the French physicist Louis Victor de Broglie (1892-1987) and Austrian physicist Erwin Schrödinger (1887-1961) that depicted the particle and wave nature of electrons. De Broglie showed that the electron was not merely a particle but a waveform.

This proposal led Schrödinger to publish his wave equation in 1926. Schrödinger's work described electrons as a "standing wave" surrounding the nucleus, and his system of quantum mechanics is called wave mechanics. German physicist Max Born (1882-1970) and English physicist P. A. M. Dirac (1902-1984) made further advances in defining the subatomic particles (principally the electron) as a wave rather than as a particle and in reconciling portions of quantum theory with relativity theory.

Working at about the same time, German physicist Werner Heisenberg (1901-1976) formulated the first complete and self-consistent theory of quantum mechanics. Matrix mathematics was well established by the 1920s, and Heisenberg applied this powerful tool to quantum mechanics.

In 1926, Heisenberg put forward his uncertainty principle which states that two complementary properties of a system, such as position and momentum, can never both be known exactly. This proposition helped cement the dual nature of particles (e.g., light can be described as having both wave and particle characteristics). Electromagnetic radiation (one region of the spectrum that comprises visible light) is now understood to have both particle and wave like properties.

In 1925, Austrian-born physicist Wolfgang Pauli (1900-1958) published the Pauli exclusion principle states that no two electrons in an atom can simultaneously occupy the same quantum state (i.e., energy state). Pauli's specification of spin (+1/2 or-1/2) on an electron gave the two electrons in any suborbital differing quantum numbers (a system used to describe the quantum state) and made completely understandable the structure of the periodic table in terms of electron configurations (i.e., the energy-related arrangement of electrons in energy shells and suborbitals).

In 1931, American chemist Linus Pauling published a paper that used quantum mechanics to explain how two electrons, from two different atoms, are shared to make a covalent bond between the two atoms. Pauling's work provided the connection needed in order to fully apply the new quantum theory to chemical reactions. Advances in nanotechnology depend upon an understanding and application of these fundamental quantum principles. At the quantum level the smoothness of classical physics disappears and nanotechnologies are predicated on exploiting this quantum roughness.

The Applications

The development of devices that are small, light, self-contained, use little energy and that will replace larger microelectronic equipment is one of the first goals of the anticipated nanotechnology revolution. The second phase will be marked by the introduction of materials not feasible at larger than nanotechnology levels. Given the nature of quantum variance, scientists theorize that single molecule sensors can be developed and that sophisticated memory storage and neural-like networks can be achieved with a very small number of molecules. Traditional engineering concepts undergo radical transformation at the atomic level. For example, nano-technology motors may drive gears, the cogs of which are composed of the atoms attached to a carbon ring. Nanomotors may themselves be driven by oscillating magnetic fields or high precision oscillating lasers.

Perhaps the greatest promise for nanotechnology lies in potential biotechnology advances. Potential nano-level manipulation of DNA offers the opportunity to radically expand the horizons of genomic medicine and immunology. Tissue-based biosensors may unobtrusively be able to monitor and regulate site-specific medicine delivery or regulate physiological processes. Nanosystems might serve as highly sensitive detectors of toxic substances or used by inspectors to detect traces of biological or chemical weapons.

In electronics and computer science, scientists assert that nanotechnologies will be the next major advance in computing and information processing science. Microelectronic devices rely on recognition and flips in electron gating (e.g. where differential states are ultimately represented by a series of binary numbers ["0" or "1"] that depict voltage states). In contrast, future quantum processing will utilise the identity of quantum states as set forth by quantum numbers. In quantum cryptography systems with the ability to decipher encrypted information will rely on precise knowledge of manipulations used to achieve various atomic states. Nanoscale devices are constructed using a combination of fabrication steps. In the initial growth stage, layers of semiconductor materials are grown on a dimension limiting substrate. Layer composition can be altered to control electrical and/or optical characteristics. Techniques such as molecular beam epitaxy (MBE) and metallo-organic chemical vapor deposition (MOCVD) are capable of producing layers of a few atoms thickness. The developed pattern is then imposed on successive layers (the pattern transfer stage) to develop desired three dimensional structural characteristics.

Nanotechnology Research

In the United States, expenditures on nanotechnology development tops $500 million per year and is largely coordinated by the National Science Foundation and Department of Defence Advanced Research Projects Agency (DARPA) under the umbrella of the National Nano-technology Initiative. Other institutions with dedicated funding for nanotechnology include the Department of Energy (DOE) and National Institutes of Health (NIH).

Research interests. Current research interests in nano-technology include programs to develop and exploit nanotubes for their ability to provide extremely strong bonds. Nanotubes can be flexed and woven into fibres for use in ultrastrong-but also ultralight-bulletproof vests. Nanotubes are also excellent conductors that can be used to develop precise electronic circuitry.

Other interests include the development of nanotechnology-based sensors that allow smarter autonomous weapons capable of a greater range of adaptations enroute to a target; materials that offer stealth characteristics across a broader span of the electromagnetic spectrum; self-repairing structures; and nanotechnology-based weapons to disrupt-but not destroy-electrical system infrastructure.

Progress of Nanotechnology

A basic question about nanotechnology is, "When will it be achieved?" The answer is simple: No one knows. How molecular machines will behave is a matter for calculation, but how long it will take us to develop them is a separate issue. Technology timetables can't be calculated from the laws of nature, they can only be guessed at. In this chapter, we examine different paths to nanotechnology, hear what some of the pioneers have to say, and describe the progress already made. This will not answer our basic question, but it will educate our guesses.

Molecular nanotechnology could be developed in any of several basically different ways. Each of these basic alternatives itself includes further alternatives. Researchers will be asking, "How can we make the fastest progress?" To understand the answers they may come to, we need to ask the same question here, adopting (for the moment) a gung-ho, let's-go, how-do-we-get-the-job-done? attitude. We give some of the researchers' answers in their own words.

Like "When will it be achieved?", this is a basic question with an answer beyond calculation. Here, though, the answer seems fairly clear. Throughout history, people have worked to achieve better control of matter, to convince atoms to do what we want them to do. This has gone on since before people learned that atoms exist, and has accelerated ever since. Although different industries use different materials and different tools and methods, the basic aim is always the same. They seek to make better things, and make them more consistently, and that means better control of the structure of matter. From this perspective, nanotechnology is just the next, natural step in a progression that has been under way for millennia.

Consider the compact discs now replacing older stereo records: both the old and the new technologies stamp patterns into plastic, but for CDs, the bumps on the stamping surface are only about 130 by 600 nanometres in size, versus 100,000 nanometres or so for the width of the groove on an old-style record. Or look at a personal computer.

John Foster, a physicist at IBM's Almaden Research Centre, points to a hard disk and says that "inside that box are a bunch of whirring disks, and every one of those disks has got a metal layer where the information is stored. The last thing on top of the metal layer is a monolayer that's the lubricant between the disk and the head that flies over it.

The monolayer is not fifteen angstroms [15 angstroms = 1.5 nanometres] and it's not three, because fifteen won't work and neither will three. So it has to be ten plus or minus a few angstroms. This is definitely working in the nanometre regime. We're at that level: We ship it every day and make money on it every day."

The transistors on computer chips are heading down in size on an exponential curve. Foster's colleague at IBM, Patrick Arnett, expects the trend to continue: "If you stay on that curve, then you end up at the atomic scale at 2020 or so. That's the nature of technology now. You expect to follow that curve as far as you can go."

The trend is clear, and at least some of the results can be foreseen, but the precise path and timetable for the development of nanotechnology is unpredictable. This unpredictability goes to the heart of important questions: "How will this technology be developed? Who will do it? Where? When? In ten years? Fifty? A hundred? Will this happen in my lifetime?" The answers will depend on what people do with their time and resources, which in turn will depend on what goals they think are most promising. Human attitudes, understanding, and goals will make all the difference.

Decisions Affecting the Rate of Advance

Decisions about research directions are central. Researchers are already pouring effort into chemical synthesis, molecular engineering, and related fields. The same amount of effort could produce more impressive results in molecular nanotechnology if a fraction of it were differently directed.

The research funders-corporate executives, and decision makers in science funding agencies like the National Science Foundation in the United States and Japan's Ministry of International Trade and Industry-all have a large influence on research directions, but so do the researchers working in the labs. They submit proposals to potential funders (and often spend time on personally chosen projects, regardless of funding), so their opinions also shape what happens. Where public money is involved, politicians' impressions of public opinion can have a huge influence, and public opinion depends on what all of us think and say..

Still, researchers play a central role. They tend to work on what they think is interesting, which depends on what they think is possible, which depends on the tools they have or-among the most creative researchers on the tools they can see how to make. Our tools shape how we think: as the saying goes, when all you have is a hammer, everything looks like a nail. New tools encourage new thoughts and enable new achievements, and decisions about tool development will pace advances in nanotechnology. To understand the challenges ahead, we need to take a look at ideas about the tools that will be needed.

Importance of Tools

Throughout history, limited tools have limited achievement. Leonardo da Vinci's sixteenth century chain drives and ball bearings were theoretically workable, yet never worked in their inventor's lifetime. Charles Babbage's nineteenth century mechanical computer suffered the same fate. The problem? Both inventors needed precisely machined parts that (though readily available today) were beyond the manufacturing technology of their times. Physicist David Miller recounts how a sophisticated integrated circuit design project at TRW hit similar limits in the early 1980s: "It all came down to whether a German company could cool their glass lenses slowly enough to give us the accuracy we needed. They couldn't."

In the molecular world, tool development again paces progress, and new tools can bring breathtaking advances. Mark Pearson, director of molecular biology for Du Pont, has observed this in action: "When I was a graduate student back in the 1950s, it was a multiyear problem to determine the molecular structure of a single protein. We used to say, 'one protein, one career.' Yet now the time has shrunk from a career to a decade to a year-and in optimal cases to a few months." Protein structures can be mapped atom by atom by studying X-ray reflections from layers in protein crystals. Pearson observes that "Characterizing a protein was a career-long endeavour in part because it was so difficult to get crystals, and just getting the material was a big constraint. With new technologies, we can get our hands on the material now-that may sound mundane, but it's a great advance.

To the people in the field, it makes all the difference in the world." Improved tools for making and studying proteins are of special importance because proteins are promising building blocks for first-generation molecular machines.

Science Discovery

Nobel Prizes are more often awarded for discoveries than for the tools (including instruments and techniques) that made them possible. If the goal is to spur scientific progress, this is a shame. This pattern of reward extends throughout science, leading to a chronic underinvestment in developing new tools. Philip Abelson, an editor of the journal Science, points out that the United States suffers from "a lack of support for development of new instrumentation. At one time, we had a virtual monopoly in pioneering advances in instrumentation. Now practically no federal funds are available to universities for the purpose." It's easier and less risky to squeeze one more piece of data out of an existing tool than to pioneer the development of a new one, and it takes less imagination.

But new tools emerge anyway, often from sources in other fields. The study of protein crystals, for example, can benefit from new X-ray sources developed by physicists, and techniques from chemistry can help make new proteins. Because they can't anticipate tools resulting from innovations in other fields, scientists and engineers are often too pessimistic about what can be achieved in their own fields. Nanotechnology will join several fields, and yield tools useful in many others. We should expect surprising results.

Researchers' Tools

Today's tools for making small-scale structures are of two kinds: molecular-processing tools and bulk-processing tools. For decades, chemists and molecular biologists have been using better and better molecular-processing tools to make and manipulate precise, molecular structures. These tools are of obvious use. Physicists, as we will see, have recently developed tools that can also manipulate molecules. Combined with techniques from chemistry and molecular biology, these physicist's tools promise great advances. Microtechnologists have applied chip-making techniques to the manufacture of microscopic machines. These technologies-the main approach to miniaturization in recent decades-can play at most a supporting role in the development of nanotechnology.

Despite appearances, it seems that microtechnology cannot be refined into nanotechnology.

Small Microtechnology

For many years, it was conventional to assume that the road to very small devices led through smaller and smaller devices: a top-down path. On this path, progress is measured by miniaturization: How small a transistor can we build? How small a motor? How thin a line can we draw on the surface of a crystal? Miniaturization focuses on scale and has paid off well, spawning industries ranging from watchmaking to microelectronics.

Researchers at AT&T Bell Labs, the University of California at Berkeley, and other laboratories in the United States have used micromachining (based on microelectronic technologies) to make tiny gears and even electric motors. Micromachining is also being pursued successfully in Japan and Germany.

These microgears and micromotors are, however, enormous by nanotechnological standards: a typical device is measured in tens of micrometres, billions of times the volume of comparable nanogears and nanomotors. (In our simulated molecular world, ten microns is the size of a small town.) In size, confusing microtechnology with molecular nanotechnology is like confusing an elephant with a ladybug.

The differences run deeper, though. Microtechnology dumps atoms on surfaces and digs them away again in bulk, with no regard for which atom goes where. Its methods are inherently crude. Molecular nanotechnology, in contrast, positions each atom with care. As Bill DeGrado, a protein chemist at Du Pont, says, "The essence of nanotechnology is that people have worked for years making things smaller and smaller until we're approaching molecular dimensions. At that point, one can't make smaller things except by starting with molecules and building them up into assemblies." The difference is basic: In microtechnology, the challenge is to build smaller; in nanotechnology, the challenge is to build bigger-we can already make small molecules.

A language warning: in recent years, nanotechnology has indeed been used to mean "very small microtechnology"; for this usage, the answer to the above question is yes, by definition.

This use of a new word for a mere extension of an old technology will produce considerable confusion, particularly in light of the widespread use of nanotechnology in the sense found here. Nanolithography, nanoelectronics, nanocomposites, nanofabrication: not all that is nano- is molecular, or very relevant to the concerns raised in this book.

The terms molecular nanotechnology and molecular manufacturing are more awkward but avoid this confusion.

Microtechnology Leading to Nanotechnology

Can bulldozers can be used to make wristwatches? At most, they can help to build factories in which watches are made.

Though there could be surprises, the relevance of microtechnology to molecular nanotechnology seems similar. Instead, a bottom-up approach is needed to accomplish engineering goals on the molecular scale.

Tools Used for Molecular Engineering

Almost by definition, the path to molecular nanotechnology must lead through molecular engineering. Working in different disciplines, driven by different goals, researchers are making progress in this field.

Chemists are developing techniques able to build precise molecular structures of sorts never before seen. Biochemists are learning to build structures of familiar kinds, such as proteins, to make new molecular objects. In a visible sense, most of the tools used by chemists and biochemists are rather unimpressive.

They work on countertops cluttered with dishes, bottles, tubes, and the like, mixing, stirring, heating, and pouring liquids-in biochemistry, the liquid is usually water with a trace of material dissolved in it.

Periodically, a bit of liquid is put into a larger machine and a strip of paper comes out with a graph printed on it. As one might guess from this description, research in the molecular sciences is usually much less expensive than research in high-energy physics (with its multibillion-dollar particle accelerators) or research in space (with its multibillion-dollar spacecraft).

Chemistry has been called "small science," and not because of the size of the molecules. Of these fields, it is biomolecular science that is most obviously developing tools that can build nanotechnology, because biomolecules already form molecular machines, including devices resembling crude assemblers.

This path is easiest to picture, and can surely work, yet there is no guarantee that it will be fastest: research groups following another path may well win. Each of these paths is being pursued worldwide, and on each, progress is accelerating.

Physicists have recently contributed new tools of great promise for molecular engineering. These are the proximal probes, including the scanning tunneling microscope (STM) and the atomic force microscope (AFM).

A proximal-probe device places a sharp tip in proximity to a surface and uses it to probe (and sometimes modify) the surface and any molecules that may be stuck to it.

Molecular Building Blocks

A huge technology base for molecular construction already exists. Tools originally developed by biochemists and biotechnologists to deal with molecular machines found in nature can be redirected to make new molecular machines. The expertise built up by chemists in more than a century of steady progress will be crucial in molecular design and construction.

Both disciplines routinely handle molecules by the billions and get them to form patterns by self-assembly. Biochemists, in particular, can begin by copying designs from nature.

Molecular building-block strategies could work together with proximal probe strategies, or could replace them, jumping directly to the construction of large numbers of molecular machines. Either way, protein molecules are likely to play a central role, as they do in nature.

6

Protein Engineering Building Molecular Machines

Proteins can self assemble into working molecular machines, objects that do something, such as cutting and splicing other molecules or making muscles contract. They also join with other molecules to form huge assemblies like the ribosome (about the size of a washing machine, in our simulation view). Ribosomes-programmable machines for manufacturing proteins-are nature's closest approach to a molecular assembler.

The genetic-engineering industry is chiefly in the business of reprogramming natural nanomachines, the ribosomes, to make new proteins or to make familiar proteins more cheaply. Designing new proteins is termed protein engineering. Since biomolecules already form such complex devices, it's easy to see that advanced protein engineering could be used to build first-generation nanomachines.

Fancy Molecular Machines

Making proteins is easier than designing them. Protein chemists began by studying proteins found in nature, but have only recently moved on to the problem of engineering new ones. These are called de novo proteins, meaning completely new, made from scratch. Designing proteins is difficult because of the way they are constructed. As Bill DeGrado, a protein chemist at Du Pont, explains: "A characteristic of proteins is that their activities depend on their three-dimensional structures. These activities may range from hormonal action to a function in digestion or in metabolism.

Whatever their function, it's always essential to have a definite three-dimensional shape or structure." This three-dimensional structure forms when a chain folds to form a compact molecular object. To get a feel for how tough it is to predict the natural folding of a protein chain, picture a straight piece of cord with hundreds of magnets and sticky knots along its length. In this state, it's easy to make and easy to understand. Now pick it up, put it in a glass jar, and shake it for a long time. Could you predict its final shape? Certainly not: it's a tangled mess. One might call this effort at prediction "the sticky-cord-folding problem"; protein chemists call theirs "the protein-folding problem."

Given the correct conditions, a protein chain always folds into one special shape, but that shape is hard to predict from just the straightened structure. Protein designers, though, face the different job of first determining a desired final shape, and then figuring out what linear sequence of amino acids to use to make that shape. Without solving the classic protein-folding problem, they have begun to solve the protein-design problem.

The Accomplishments so Far

Bill DeGrado and his colleagues at Du Pont had one of the first successes: "We've been able to use basic principles to design and build a simple molecule that folds up the way we want it to. This is really the first real example of a designed protein structure, designed from scratch, not by taking an already existing structure and tinkering with it."

Although scientists do the work, the work itself is really a form of engineering, as shown by the title of the field's journal, Protein Engineering. Bill DeGrado's description of the process makes this clear: "After you've made it, the next step is to find out whether your protein did what you expected it to do. Did it fold? Did it pass ions across bilayers [such as cell membranes]? Does it have a catalytic function [speeding specific chemical reactions]? And that's tested using the appropriate experiment. More than likely, it won't have done what you wanted it to do, so you have to find out why. Now, a good design has in it a contingency plan for failure and helps you learn from mistakes. Rather than designing a structure that would take a year or more to analyze, you design it so that it can be assayed for given function or structure in a matter of days."

Many groups are pursuing protein design today, including academic researchers like Jane and Dave Richardson at Duke University, Bruce Erickson at the University of North Carolina, and Tom Blundell, Robin Leatherbarrow, and Alan Fersht in Britain. The successes have started to roll in. Japan, however, is unique in having an organisation devoted exclusively to such projects: the Protein Engineering Research Institute (PERI) in Osaka. In 1990,

PERI announced the successful design and construction of a de novo protein several times larger than any built before.

Speciality of Proteins

The main advantage of proteins is that they are familiar: a lot is known about them, and many tools exist for working with them. Yet proteins have disadvantages as well. Just because this design work is starting with proteins-soft, squishy molecules that are only marginally suitable for nanotechnology-doesn't mean it will stay within those limits.

De Grado points out "The fundamental goal of our work in de novo design is to be able to take the next step and get entirely away from protein systems." An early example is the work of Wallace Carothers of Du Pont, who used a de novo approach to studying the nature of proteins: Rather than trying to cut up proteins, he tried to build up things starting with amino acids and other similar monomers. In 1935, he succeeded in making nylon.

DeGrado explains "There is a deep philosophical belief at Du Pont in the ability of people to make molecules de novo that will do useful things. And there is a fair degree of commitment from the management that following that path will lead to products: not directly, and not always predictably, but they know that they need to support the basic science.

"I think ultimately we have a better chance at doing some really exciting things by de novo design, because our repertory should be much greater than that of nature. Think about the ability to fly: One could breed better carrier pigeons or one could design airplanes."

The biology community, however, leans more toward ornithology than toward aerospace engineering. DeGrado's experience is that "a lot of biologists feel that if you aren't working with the real thing [natural proteins], you aren't studying biology, so they don't totally accept what we're doing. On the other hand, they recognise it as good chemistry."

Where is Protein Engineering Headed?

Like the IBM physicists, protein designers are moved by a vision of molecular engineering. In 1989, Bill DeGrado predicted, "I think we'll be able to make catalysts or enzymelike molecules, possibly ones that catalyze reactions not catalyzed in nature." Catalysts are molecular machines that speed up chemical reactions: they form a shape for the two reacting molecules to fit into and thereby help the reaction move faster, up to a million reactions per second. New ones, for reactions that now go slowly, will give enormous cost savings to the chemical industry.

This prediction was borne out just a few months later, when Denver researchers John Stewart, Karl Hahn, and Wieslaw Klis announced their new enzyme, designed from scratch over a period of two years and built successfully on the first try. It's a catalyst, making some reactions go about 100,000 times faster. Nobel Prize-winning biochemist Bruce Merrifield believes that "if others can reproduce and expand on this work, it will be one of the most important achievements in biology or chemistry."

DeGrado also has longer term plans for protein design, beyond making new catalysts: "It will allow us to think about designing molecular devices in the next five to ten years. It should be possible ultimately to specify a particular design and build it. Then you'll have, say, proteinlike molecules that self-assemble into complex molecular objects, which can serve as machinery. But there's a limit to how small you can make devices. You'll shrink things down so far and then you won't be able to go any further, because you've reached molecular dimensions."

Mark Pearson shows that management at Du Pont also has this vision. Regarding the prospects for nanotechnology and assemblers, he remarked, "You know, it'll take money and effort and good ideas for sure. But to my way of thinking, there is no absolute fundamental limitation to preclude us from doing this kind of thing." He didn't say his company plans to develop nanotechnology, but such plans aren't really necessary. Du Pont is already on the nanotechnology path, for other-shorter-term, commercial-reasons. Like IBM, if they do decide to move quickly, they have the resources and forward-looking people needed to succeed.

Who Else Builds Molecular Objects?

Chemists, most of whom do not work on proteins, are the traditional experts in building molecular objects. As a group they've been building molecules for over a century, with ever increasing ability and confidence. Their methods are all indirect: They work with billions of atoms at a time-massive parallelism-but without control of the positions of their workpieces. The molecules typically tumble randomly in a liquid or gas, like pieces of a puzzle that may or may not fit together correctly when shaken together in a box. With clever design and planning, most pieces will join properly.

Chemists mix molecules on a huge scale (in our simulation view, a test tube holds a churning molecular swarm with the volume of an inland sea), yet they still achieve precise molecular transformations. Given that they work so indirectly, their achievements are astounding. This is, in part, the result of the enormous amount of work poured into the field for many decades. Thousands of chemists are working on molecular construction in the United States alone; add to that the chemists in Europe, in Japan, and in the rest of the world, and you have a huge community of researchers making great strides. Though it publishes only a one-paragraph summary of each research report, a guide to the chemical literature-Chemical Abstracts-covers several library walls and grows by many feet of shelf space every year.

How can Mixing Chemicals Build Molecular Objects?

An engineer would say that chemists (at least those specialising in synthesis) are doing construction work, and would be amazed that they can accomplish anything without being able to grab parts and put them in place. Chemists, in effect, work with their hands tied behind their backs.

Molecular manufacturing can be termed "positional chemistry" or "positional synthesis," and will give chemists the ability to put molecules where they want them in three-dimensional space. Rather than trying to design puzzle pieces that will stick together properly by themselves when shaken together in a box, chemists will then be able to treat molecules more like bricks to be stacked. The basic principles of chemistry will be the same, but strategies for construction will become far simpler.

Without positional control, chemists face a problem something like this: Picture a giant glass barrel full of tiny battery-powered drills, buzzing away in all directions, vibrating around in the barrel. Your goal is to take a piece of wood and put a hole in just one specific spot. If you simply throw it in the barrel, it will be drilled haphazardly in many places.

To control the process, you must protect all the places you don't want drilled-perhaps by gluing protective pieces of metal over most of the wood surface. This problem-how to protect one part of a molecule while altering another part-has forced chemists to develop ever-cleverer ploys to build larger and larger molecules.

If Chemists can Make Molecules, Why aren't They Building Fancy Molecular Machines?

Chemists can achieve great things, but have focused much of their effort on duplicating molecules found in nature and then making minor variants. As an example, take palytoxin, a molecule found in a Hawaiian coral. It was so difficult to make in the lab that it has been called "the Mount Everest of synthetic chemistry," and its synthesis was hailed as a triumph. Other efforts are poured into making small molecules with unusual bonding, or molecules of remarkable symmetry, like "cubane" and "dodecahedrane" (shaped like the Platonic solids they are named after).

Chemists, at least in the United States, regard themselves as natural scientists even when their life's work is the construction of molecules by artificial means. Ordinarily, people who build things are called engineers. And indeed, at the University of Tokyo the Department of Synthetic Chemistry is part of the Faculty of Engineering; its chemists are designing molecular switches for storing computer data. Engineering achievements will require work directed at engineering goals.

Building Molecular Machines

Molecular engineers working toward nanotechnology need a set of molecular building blocks for making large, complex structures. Systematic building-block construction was pioneered by Bruce Merrifield, winner of the 1984 Nobel Prize in Chemistry. His approach, known as "solid phase synthesis," or simply "the Merrifield method," is used to synthesize the long chains of amino acids that form proteins.

In the Merrifield method, cycles of chemical reaction each add one molecular building block to the end of a chain anchored to a solid support. This happens in parallel to each of trillions of identical chains, building up trillions of molecular objects with a particular sequence of building blocks. Chemists routinely use the Merrifield method to make molecules larger than palytoxin, and related techniques are used for making DNA in so-called gene machines: an ad from an Alabama company reads, "Custom DNA-Purified and Delivered in 48 hours."

While it's hard to predict how a natural protein chain will fold-they weren't designed to fold predictably-chemists could make building blocks that are larger, more diverse, and more inclined to fold up in a single, obvious, stable pattern. With a set of building blocks like these, and the Merrifield method to string them together, molecular engineers could design and build molecular machines with greater ease.

How do Researchers Design What They can't See?

To make a new molecule, both its structure and the procedure to make it must be designed. Compared to gigantic science projects like the Superconducting Supercollider and the Hubble Space Telescope, working with molecules can be done on a shoestring budget. Still, the costs of trying many different procedures add up. To help predict in advance what will work and what won't, designers turn to models.

You may have played with molecular models in chemistry class: coloured plastic balls and sticks that fit together like Tinker Toys. Each colour represents a different kind of atom: carbon, hydrogen, and so on. Even simple plastic models can give you a feel for how many bonds each kind of atom makes, how long the bonds are, and at what angles they are made. A more sophisticated form of model uses only spheres and partial spheres, without sticks.

These colourful, bumpy shapes are called CPK models, and are widely used by professional chemists. Nobel laureate Donald Cram remarks that "We have spent hundreds of hours building CPK models of potential complexes and grading them for desirability as research targets." His research, like that of fellow Nobelists Charles J. Pedersen and Jean-Marie Lehn, has focused on designing and making medium-sized molecules that self assemble.

Although physical models can't give a good description of how molecules bend and move, computer-based molecules can. Computer-based modelling is already playing a key role in molecular engineering.

As John Walker (a founder and leader of Autodesk) has remarked, "Unlike all of the industrial revolutions that preceded it, molecular engineering requires, as an essential component, the ability to design, model, and simulate molecular structures using computers."

This has not gone unnoticed in the business community. John Walker's remark was part of a talk on nanotechnology given at Autodesk, a leader in computer-aided design and one of the five largest software firms in the United States. Soon after this talk, the company made its first major investment in the computer-aided design of molecules.

More Familiar Kinds of Engineering

Manufacturers and architects know that designs for new products and buildings are best done on a computer, by computer-aided design (CAD). The new molecular design software can be called molecular CAD, and in its forefront are researchers such as Jay Ponder of the Yale University Department of Molecular Biophysics and Biochemistry. Ponder explains that "There's a strong link between what molecular designers are doing and what architects do. Michael Ward of Du Pont is designing a set of building blocks for a Tinker Toy set so that you can build larger structures. That's exactly what we're doing with molecular modelling techniques.

"All the design and mechanical engineering principles that apply to building a skyscraper or a bridge apply to molecular architecture as well. If you're building a bridge, you're going to model it and see how many trucks can be on the bridge at the same time without it collapsing, what kind of forces you're going to apply to it, whether it can stand up to an earthquake.

"And the same process goes on in molecular design: You're designing pieces and then analyzing the stresses and forces and how they will change and perturb the structure. It's exactly the same as designing and building a building, or analyzing the stresses on any macroscale structure. I think it's important to get people to think in those terms.

"The molecular designer has to be creative in the same way that an architect has to be creative in designing a building. When people are looking at the interior of a protein structure and trying to redesign it to create a space that will have a particular function, such as binding to particular molecules, that's like designing a room to use as a dining room-one that will fit certain sizes of tables and certain numbers of guests. It's the same thing in both cases: You have to design a space for a function."

Ponder combines chemistry and computer science with an overall engineering approach: "I'm kind of a hybrid. I spend about half my time doing experiments and about half my time writing computer programs and doing computational work. In the laboratory, I create or design molecules to test some of the computational ideas. So I'm at the interface." The engineering perspective helps in thinking about where molecular research can lead: "Even though with nanotechnology we're at the nanometre scale, the structures are still big enough that an awful lot of things are classical. Again, it's really like building bridges-very small bridges. And so there are many almost standard mechanical-engineering techniques for architecture and building structures, such as stress analysis, that apply."

Getting to nanotechnology will require the work of experts in differing fields: chemists, who are learning how to make molecular machines; computer scientists, who are building the needed design tools; and perhaps STM and AFM experts, who can provide tools for molecular positioning. To make progress, however, these experts must do more than just work, they must work together. Because nanotechnology is inherently interdisciplinary, countries that draw hard lines between their academic disciplines, as the United States does, will find that their researchers have difficulty communicating and cooperating.

In chemistry today, a half-dozen researchers aided by a few tens of students and technicians is considered a large team. In aerospace engineering, enormous tasks like reaching the Moon or building a new airliner are broken down into tasks that are within the reach of small teams. All these small teams work together, forming a large team that may consist of thousands of engineers aided by many thousands of technicians. If chemistry is to move in the direction of molecular systems engineering, chemists will need to take at least a few steps in this direction.

In engineering, everyone knows that designing a rocket will require skills from many disciplines. Some engineers know structures, others know pumps, combustion, electronics, software, aerodynamics, control theory, and so on and so forth down a long list of disciplines. Engineering managers know how to bring different disciplines together to build systems.

In academic science, interdisciplinary work is productive and praised, but is relatively rare. Scientists don't need to cooperate to have their results fit together: they are all describing different parts of the same thing-nature-so in the long run, their results tend to come together into a single picture. Engineering, however, is different. Because it is more creative (it actually creates complex things), it demands more attention to teamwork. If the finished parts are going to work together, they must be developed by groups that share a common picture of what each part must accomplish. Engineers in different disciplines are forced to communicate; the challenge of management and team-building is to make that communication happen. This will apply to engineering molecular systems as much as it does to engineering computers, cars, aircraft, or factories.

Jay Ponder suggests that it's a question of perspective. "It's all a matter of what's perceived to be important by the different groups that have to come together to make this work: the chemists doing their bit and the computational people doing their bit. People have to come together and see the big picture. There are people who try to bridge the gaps, but they are rare compared to the people who just work in their own speciality." Progress toward nanotechnology will continue, and as it does, researchers trained as chemists, physicists, and the like will learn to talk to one another to solve new problems. They will either learn to think like engineers and work in teams, or they will be eclipsed by colleagues who do.

With all these problems, the advance toward nanotechnology steadily continues. Industry must gain ever-better control of matter to stay competitive in the world marketplace. The STM, protein engineering, and much of chemistry are driven by commercial imperatives. Focused efforts would yield faster advances, yet even without a clear focus, advances in this direction have an air of inevitability. As Bill DeGrado observes, "We really do have the tools. Experience has shown that when you have the analytic and synthetic tools to do things, in the end science goes ahead and does them-because they are doable."

Jay Ponder agrees: "Over the next few years, you're going to see slow evolutionary advances coming from people tinkering with molecular structures and figuring out their principles. People are going to work on a particular problem because they see some application for it or because they got grant funding for it. And in the process of doing something like improving a laundry detergent's ability to clean protein stains, Proctor and Gamble is going to help work out the principles for how to increase molecular stability, and to design spaces inside the molecules."

For a variety of reasons, Japan's contribution to nanotechnology research promises to be excellent. While the United States has generally pursued researching this area with little sense of long-term direction, it appears that Japan has begun to take a more focused approach. Researchers there already have clear ideas about molecular machines-about what might work and what probably won't. Japanese researchers are accustomed to a higher level of interdisciplinary contact and engineering emphasis than are Americans. In the

United States, we prize "basic science," often calling it "pure science," as if to imply that practical applications are a form of impurity. Japan instead emphasizes "basic technology."

Nanotechnology is a basic technology, and the Japanese recognise it as such. Recent changes at the Tokyo Institute of Technology-Japan's equivalent of MIT-reflect their views of promising directions for future research. For many decades, Tokyo Tech has had two major divisions: a Faculty of Science and a Faculty of Engineering. To these is now being added a Faculty of Bioscience and Biotechnology, to consist of four departments: a Department of Bioscience, a Department of Bioengineering, a Department of Biomolecular Engineering, and what is termed a "Department of Biostructure." The creation of a new faculty in a major Japanese university is a rare event. What U.S university has a department explicitly devoted to molecular engineering? Japan has both the departments at Tokyo Tech and Kyoto University's recently established Department of Molecular Engineering.

Japan's Institute for Physical and Chemical Research (RIKEN) has broad-based interdisciplinary strength. Hiroyuki Sasabe, head of the Frontier Materials Research Program at RIKEN, notes that the institute has expertise in organic synthesis, protein engineering, and STM technology. Sasabe says that his laboratory may need a molecular manipulator of the sort described in the next chapter to accomplish its goals in molecular engineering. Research consortia in Japan are also moving toward nanotechnology. The Exploratory Research for Advanced Technology Organisation (ERATO) sponsors many three-to-five year projects in parallel, each with a specific goal. Consider the work in progress:

- Yoshida Nanomechanism Project
- Hotani Molecular Dynamic Assembly Project
- Kunitake Molecular Architecture Project
- Nagayama Protein Array Project
- Aono Atomcraft Project

These focus on different aspects of gaining control over matter at the atomic level. The Nagayama Protein Array Project aims to use proteins as engineering materials to move toward making new molecular devices. The Aono Atomcraft Project does not involve nuclear power-as its translation might imply-but is instead an interdisciplinary effort to use an STM to arrange matter on the atomic scale.

At some point, work on nanotechnology must move beyond spin-offs from other fields and undertake the design and construction of molecular machinery. This shift from opportunistic science to organised engineering requires a change in attitude. In this, Japan leads the United States.

Molecular nanotechnology will emerge step by step. Major milestones, such as the engineering of proteins and the positioning of individual atoms, have already been passed. To get a sense of the likely pace of developments, we need to look at how various trends fit together.

Computer-based molecular-modelling tools are spawning computer-aided design tools. These will grow more capable. The underlying technology base-computer hardware-has for decades been improving in price and performance on a steeply rising curve, which is generally expected to continue for many years. These advances are quite independent of progress in molecular engineering, but they make molecular engineering easier, speeding advances. Computer models of molecular machines are beginning to appear, and these will whet the appetites of researchers.

Progress in engineering molecular machines, whether using proximal probes or self-assembly, will eventually achieve striking successes; the objectives of research in Japan will begin to draw serious attention; understanding of the long-term promise of molecular engineering will become more widespread.

Some combination of these developments will eventually lead to a serious, public appraisal of what these technologies can achieve-and then the world of opinion, funding, and research fashion will change. Before, advances will be steady but haphazard; afterward, advances will be driven with the energy that flows into major commercial, military, and medical research programs, because nanotechnology will be recognised as furthering major commercial, military, and medical goals. The timing of subsequent events depends largely on when this threshold of serious attention is reached.

In making time estimates, people are prone to assume that a large change must take a long time. Most do, but not all. Pocket calculators had a dramatic effect on the slide-rule industry: they replaced it.

The speed of this change caught the slide rule moguls by surprise, but the pace of progress in electronics didn't slow down merely to suit their expectations. One can argue that nanotechnology will be developed fast: many countries and companies will be competing to get there first. They will be driven onward both by the immense expected benefits-in many areas, including medicine and the environment-as well as by potential military applications. That is a powerful combination of motives, and competition is a powerful accelerator.

A counterargument, though, suggests that development will be slow: anyone who has done anything of significance in the real world of technology-doing a scientific experiment, writing a computer program, bringing a new product to market-knows that these goals take longer than expected. Indeed, Hofstadter's Law states that projects take longer than expected, even when

Hofstadter's Law is taken into account. This principle is a good guide for the short term, and for a single project.

The situation differs, though, when many different approaches are being explored by many different groups over a period of years. Most projects may take longer than expected, but with many teams trying many approaches, one approach may prove faster than expected. The winner of a race is always faster than the average runner. John Walker notes, "The remarkable thing about molecular engineering is that it looks like there are many different ways to get there and, at the moment, rapid progress is being made along every path-all at the same time."

Also, technology development is like a race run over an unmapped course. When the first runners reach the top of a hill, they may see a shortcut. A trailing runner may decide to crash off into the bushes, and stumble across a bicycle and a paved road. The progress of technology is seldom predictable because progress often reveals new directions. How close we are to goal depends on whether technological advances are a constant pace of accelerating. In this diagram, the dashed line represents the current level of technology, and the large dot in the upper right represents a goal such as nanotechnology. With a straight-line advance, it's easier to estimate how far away a goal is. With an accelerating advance, a goal can be reached with little warning.

So how can we estimate a date for the arrival of nanotechnology? It's safest to take a cautious approach: When anticipating benefits, assume it's far off; when preparing for potential problems, assume it's right around the corner. The old folk saying applies: Hope for the best, prepare for the worst. Any dates assigned to "far off" and "right around the corner" can be no better than educated guesses-molecular behaviour can be calculated, but not technology timetables of this sort. With those caveats, we would estimate that general-purpose molecular assemblers will likely be developed in the early decades of the twenty-first century, perhaps in the first.

John Walker, whose technological foresight has led Autodesk from start-up to a dominant role in its industry, points out that not long ago, "Many visionaries intimately familiar with the developments of silicon technology still forecast it would take between twenty and fifty years before molecular engineering became a reality. This is well beyond the planning horizon of most companies. But recently, everything has begun to change." Based on the new developments, Walker places his bet: "Current progress suggests the revolution may happen within this decade, perhaps starting within five years.

Behind the Scenes and Afterward

Desert Rose Industries and other manufacturers can make almost anything quickly and at low cost. That includes the tunneling machines and other equipment that made the subway system they use for shipping. Digging a tunnel from coast to coast now costs less than digging a single block under New York City used to. It wasn't expensive to get a deep-transit terminal installed in their basement. Just as the tents aren't mere bundles of canvas, these subways aren't slow things full of screeching, jolting metal boxes. They're magnetically levitated to reach aircraft speeds-as experimental Japanese trains were in the late 1980s-making it easy for Carl and Maria to give their customers quick service. There's still a road leading to the plant, but nobody's driven a truck over it for years.

They only take in materials that they will eventually ship out in products, so there's nothing left over, and no wastes to dump. One corner of the plant is full of recycling equipment. There are always some obsolete parts to get rid of, or things that have been damaged and need to be reworked. These get broken down into simpler molecules and put back together again to make new parts.

The gunk in the manufacturing ponds is water mixed with particles much finer than silt. The particles-fasteners, computers, and the rest-stay in suspension because they are wrapped in molecular jackets that keep them there. This uses the same principle as detergent molecules, which coat particles of oily dirt to float them away.

Though it wouldn't be nutritious or appetizing, you could drink the tent mix and be no worse for it. To your body, the parts and their jackets, and even the nanomachines, would be like so many bits of grit and sawdust. (Grandma would have called it roughage.)

Carl and Maria get their power from solar cells in the road, which is the only reason they bothered having it paved. In back of their plant stands what looks like a fat smokestack. All it produces, though, is an updraft of clean, warm air. The darkly paved road, baking in the New Mexico sun, is cooler than you might expect: it soaks up solar energy and makes electricity, instead of just heat. Once the power is used, it turns back into heat, which has to go somewhere. So the heat rises from their cooling tower instead of the road, and the energy does useful work on the way.

Some products, like rocket engines, are made more slowly and in a single piece. This makes them stronger and more permanent. The tents, though, don't need to be superstrong and are just for temporary use. A few days after the tents go up, the earthquake victims start to move out into new housing (permanent, better-looking, and very earthquake resistant). The tents get folded and shipped off for recycling.

Recycling things built this way is simple and efficient: nanomachines just unscrew and unsnap the connectors and sort the parts into bins again. The shipments Desert Rose gets are mostly recycled to begin with. There's no special labelling for recycled materials, because the molecular parts are the same either way.

For convenience (and to keep the plant small), Carl and Maria get most of their parts prefabricated, even though they can make almost anything. They can even make more production equipment. In one of their manufacturing ponds, they can put together a new cabinet full of special-purpose assemblers. They do this when they want to make a new type of part in-house. Like parts, the part-assemblers are made by special-purpose assemblers. Carl can even make big vats in medium-size vats, unfolding them like tents.

If Desert Rose Industries needed to double capacity, Carl and Maria could do it in just a few days. They did this once for a special order of stadium sections. Maria got Carl to recycle the new building before its shadow hurt their cactus garden.

In the Desert Rose Industries scenario, manufacturing has become cheap, fast, clean, and efficient. Using fast, precise machines to handle matter in molecular pieces makes it easy for nanotechnology to be fast, clean, and efficient. But for it to be cheap, the manufacturing equipment has to be cheap. The Desert Rose scenario shows how this can work. Molecular-manufacturing equipment can be used to make all the parts needed to build more molecular manufacturing equipment. It can even build the machines needed to put the parts together. This resembles an idea developed by NASA for a self-expanding manufacturing complex on the Moon, but made faster and simpler using molecular machines and parts.

In the early days of nanotechnology, there won't be as many different kinds of machines as there are at Desert Rose. One way to build a lot of molecular manufacturing equipment in a reasonable time would be to make a machine that can be used to make a copy of itself, starting with special but simple chemicals. A machine able to do this is called a "replicator." With a replicator and a pot full of the right fuel and raw materials, you could start with one machine, then have two, four, eight, and so on.

This doubling process soon makes enough machines to be useful. The replicators-each including a computer to control it and a general-purpose assembler to build things-could then be used to make something else, like the tons of specialised machines needed to set up a Desert Rose manufacturing plant. At that point, the replicators could be discarded in favour of those more efficient machines.

Replicators are worth a closer look, though, because they show how quickly molecular manufacturing systems can be used to build more manufacturing equipment. If we were in one of our standard simulation views, the submicroscopic device at the top of the picture would be like a huge tank, three stories tall when lying on its side. Most of its interior is taken up by a tape memory system that tells how to move the arm to build all the parts of the replicator, except the tape itself. The tape gets made by a special tape-copying machine. At the right-hand end of the replicator are pores for bringing in fuel and raw-material molecules, and machinery for processing them. In the middle are computer-controlled arms, like the ones we saw on the plant trip. These do most of the actual construction.

The steps in the cycle-using a copy to block the tube, beginning a fresh copy, then releasing the old one-illustrate one way for a machine to build a copy of itself while floating in a liquid, yet doing all its construction work inside, in vacuum. (It's easier to design for vacuum, and this is exploratory-engineering work, so easier design is better design.) Calculations suggest that the whole construction cycle can be completed in less than a quarter hour, since the replicator contains about a billion atoms, and each arm can handle about a million atoms per second. At that rate, one device can double and double again to make trillions in about ten hours.

Each replicator just sits in a chemical bath, soaking up what it needs and making more replicators. Eventually, either the special chemicals run out or other chemicals are added to signal them to do something else. At that point, they can be reprogrammed to produce anything else you please, so long as it can be extruded from the front. The products can be long, and can unfold or be pieced together to make larger objects, so the size of these initial replicators-smaller than a bacterium-would be only a temporary limitation.

From the molecular manipulators and primitive assemblers described in the last chapter, the most likely path to nanotechnology leads to assemblers with more and more general capabilities. Still, efficiency favours special-purpose machines, and the Desert Rose scenario didn't make much use of general assemblers. Why bother making general-purpose assemblers in the first place?

To see the answer, turn the question around and ask, Why not build such a tool? There is nothing outstandingly difficult about a general assembler, as molecular machinery goes. It will just be a device with good, flexible positional control and a system to feed it a variety of molecular tools. This is a useful, basic capability. General-purpose assemblers could always be replaced by a lot of specialised devices, but to build those specialised devices in the first place, it makes sense to come up with a more flexible, general-purpose system that can just be reprogrammed.

So, general purpose machines are likely to find use in making short production runs of more specialised devices. Ralph Merkle, a computers and security expert at Xerox Palo Alto Research Centre, sees this as paralleling the way manufacturing works today: "General purpose devices could do many tasks, but they'll do them inefficiently. For any given task, there will be one or a few best ways of doing it, and one or a few special purpose devices that are finely tuned to do that one task. Nails aren't made by a general-purpose machine shop, they're made by nail-making machines.

Making nails with a general-purpose machine shop would be more expensive, more difficult, and more time-consuming. Likewise, in the future we won't see a proliferation of general-purpose self-replicating systems, we'll see specialisation for almost every task." We've surveyed a lot of devices: assemblers of various flavours, nanocomputers, disassemblers, replicators, and others. What's important about these is not the exact distinctions between them, but the capabilities that they will give and the effects they will have on human lives. Again, we are suspending discussion of potential misapplications until later.

If we tease apart the implications of what we've seen in the Desert Rose scenario, we can analyze some of the key impacts of molecular manufacturing in industry, science, and medicine.

The Industry and Technology

At its base, nanotechnology is about molecular manufacturing, and manufacturing is the basis of much of today's industry. This is why Desert Rose made a good starting point for describing the possibilities of a nanotechnological world. From an industrial perspective, it makes sense to think of nanotechnology in terms of products and production.

New Products: Today, we handle matter crudely, but nanotechnology will bring thorough control of the structure of matter, the ability to build objects to atom-by-atom specifications. This means being able to make almost anything. By comparison, even today's range of products will feel very limited. Nanotechnology will make possible a huge range of new products, a range we can't envision today. Still, to get a feel for what is possible, we can look at some easily imagined applications.

Reliable Products: Today, products often fail, but for failures to occur-for a wing to fall off an airplane, or a bearing to wear out-a lot of atoms have to be out of place. In the future, we can do better. There are two basic reasons for this: better materials and better quality control, both achieved by molecular manufacturing. By using materials tens of times stronger than steel, as Desert Rose did, it will be easy to make things that are very strong, with a huge safety margin. By building things with atom-by-atom control, flaws can be made very rare and extremely small-nonexistent, by present standards.

With nanotechnology, we can design in big safety margins and then manufacture the design with near-perfection. The result will be products that are tough and reliable. (There will still be room for bad designs, and for people who wish to take risks in machines that balance on the edge of disaster.)

Intelligent Products: Today, we make most things from big chunks of metal, wood, plastic, and the like, or from tangles of fibres. Objects made with molecular manufacturing can contain trillions of microscopic motors and computers, forming parts that work together to do something useful.

A climber's rope can be made of fibres that slide around and reweave to eliminate frayed spots. Tents can be made of parts that slide and lock to turn a package into a building. Walls and furniture can be made to repair themselves, instead of passively deteriorating.

On a mundane level, this sort of flexibility will increase reliability and durability. Beyond this, it will make possible new products with abilities we never imagined we needed so badly. And beyond even this, it will open new possibilities for art.

Inexpensive Production: Today, production requires a lot of labour, either for making things or for building and maintaining machines that make things. Labour is expensive, and expensive machines make automation expensive, too. In the Desert Rose scenario, we got a glimpse of how molecular manufacturing can make production far less expensive than it is today. This is perhaps the most surprising conclusion about nanotechnology, so we'll take a closer look at it in the next chapter.

Clean Production: Today, our manufacturing processes handle matter sloppily, producing pollution. One step puts stuff where it shouldn't be; the next washes it off the product and into the water supply. Our transportation system worsens the problem as unreliable trucks and tankers spill noxious chemicals over the land and sea. Everything is expensive, so companies skimp on even the half-effective pollution controls that we know how to build.

Nanotechnology will mean greater control of matter, making it easy to avoid pollution. This means that a little public pressure will go a long way toward a cleaner environment. Likewise, it will make it easy to increase efficiency and reduce resource requirements. Products, like the Red Cross tents at Desert Rose, can be made of snap-together, easily recyclable parts. Sophisticated products could even be made from biodegradable materials. Nanotechnology will make it easy to attack the causes of pollution at their technological root.

Nanotechnology will have great applications in the field of industry, much as transistors had great applications in the field of vacuum tube electronics, and democracy had great applications in the field of monarchy. It will not so much advance twentieth-century industry as replace it-not all at once, but during a thin slice of historical time.

Chemistry: Today, chemists work with huge number of molecules and study them using clever, indirect techniques. Making a new molecule can be a major project, and studying it can be another. Molecular manufacturing will help chemists make what they want to study, and it will help them make the tools they need to study it. Nanoinstruments will be used to prod, measure, and modify molecules in a host of ways, studying their structures, behaviours, and interactions.

Materials: Today, materials scientists make new superconductors, semiconductors, and structural materials by mixing and crushing and baking and freezing, and so forth. They dream of far more structures than they can make, and they stumble across more things than they plan. With molecular manufacturing, materials science can be much more systematic and thorough. New ideas can be tested because new materials can be built according to plan (rather than playing around, groping for a recipe).

This need not rule out unexpected discoveries, since experiments-even blind searches-will go much faster. A few tons of raw materials would be enough to make a billion samples, each a cubic micron in size. In all of history so far, materials scientists have never tested so many materials. With nanoinstruments and nanocomputers, they could. One laboratory could then do more than all of today's materials scientists put together.

Biology: Today, biologists use a host of molecular devices borrowed from biology to study biology. Many of these can be viewed as molecular machines. Nanotechnology will greatly advance biology by providing better molecular devices, better nanoinstruments. Some cells have already been mapped in amazing molecular detail, but biology still has far to go. With nanoinstruments (including molecule-by-molecule disassemblers), biologists will at last be able to map cells completely and study their interactions in detail. It will become easy not only to find molecules in cells, but to learn what they do. This will help in understanding disease and the molecular requirements for health, enormously advancing medicine.

Computation: Today, computers range from a million to a billion times faster than an old desktop adding machine, and the results have been revolutionary for science. Every year, more questions can be answered by calculations based on known principles of physics. The advent of nanocomputers-even slow, miserable, mechanical nanocomputers-will give us practical machines with a trillion times the power of today's computers (essentially by letting us package a trillion computers in a small space, without gobbling too much money or energy.) The consequences will again be revolutionary.

Physics: The known principles of physics are adequate for understanding molecules, materials, and cells, but not for understanding phenomena on a scale that would still be submicroscopic if atoms were the size of marbles. Nanotechnology can't help here directly, but it can provide manufacturing facilities that will make huge particle accelerators economical, where today they strain national budgets.

More generally, nanotechnology will help science wherever precision and fine details are important. Science frequently proceeds by trying small variations in almost identical experiments, comparing the results. This will be easier when molecular manufacturing can make two objects that are identical, molecule by molecule. In some areas, today's techniques are not only crude, but destructive. Archaeological sites are unique records of the human past, but today's techniques throw away most information during the dig, by accident. Future archaeologists, able to sift soil not speck by speck but molecule by molecule, will be grateful indeed to those archaeologists who today leave some ground undisturbed.

Of all the areas where the ability to manufacture new tools is important, medicine is perhaps the greatest. The human body is intricate, and that intricacy extends beyond the range of human vision, beyond microscopic imaging, down to the molecular scale.

"Molecular medicine" is an increasingly popular term today, but medicine today has only the simplest of molecular tools. As biology uses nanoinstruments to learn about disease and health, we will learn the physical requirements for restoring and maintaining health. And with this knowledge will come the tools needed to satisfy those requirements-tools ranging from improved pharmaceuticals to devices able to repair cells and tissues through molecular surgery.

Advanced medicine will be among the most complex and difficult applications of nanotechnology. It will require great knowledge, but nanoinstruments will help gather this knowledge. It will pose great engineering challenges, but computers of trillionfold greater power will help meet those challenges. It will solve medical problems on which we spend billions of dollars today, in hopes of modest improvements.

Today, modern medicine often means an expensive way to prolong misery. Will nanomedicine be more of the same? Any reader over the age of, say, thirty knows how things start to go wrong: an ache here, a wrinkle there, the loss of an ability. Over the decades, the physical quality of life declines faster and faster-the limits of what the body can do become stricter-until the limits are those of a hospital bed.

The healing abilities we have when young seem to fade away. Modern medical practice expends the bulk of its effort on such things as intensive care units, dragging out the last few years of life without restoring health.

Truly advanced medicine will be able to restore and supplement the youthful ability to heal. Its cost will depend on the cost of producing things more intricate than any we have seen before, the cost of producing computers, sensors, and the like by the trillions. To understand the prospects for medicine, like those for science and industry, we need to take a closer look at the cost of molecular manufacturing.

In earlier chapters, we have stepped forward and backward through time. The last step was a big one, leaping from small laboratory devices to the high-capacity industrial facility of the Desert Rose scenario. Our narrative crossed this gap in a single leap, but the world won't. To understand how nanotechnology might unfold, it makes sense to look at some of its easier and more difficult applications. The result won't be a timetable, or even a series of milestones, but it should give a better picture of what we can expect as nanotechnology develops from simple, crude, costly beginnings to a state of greater sophistication and lower cost.

Molecular manufacturing will make better products possible. We're likely to see some early applications in at least two areas: stronger materials and faster computers. Strong materials are simple, and will be hard to pass up. Computers are more complex, but the payoff will be enormous.

Strong and Lightweight Structures

At the opposite extreme from molecular electronics-complex and at first worth billions of dollars per gram-are structural materials: worth only dollars per kilogram in most applications, but much simpler in structure. Once molecular manufacturing becomes inexpensive, structural materials will be important products. These materials play a central role in almost everything around us, from cars and aircraft to furniture and houses. All of these objects get their size, shape, and strength from a structural skeleton of some sort. This makes structural materials a natural place to begin in understanding how nanotechnology can improve products.

Cars today are mostly made of steel, aircraft of aluminum, and buildings and furniture largely of steel and wood. These materials have a certain "strength-to-weight ratio" (more properly, a strength-to-density ratio). To make cars stronger, they'd have to be heavier; to make them lighter, they'd have to be weaker. Clever design can change this relationship a little, but to change it a lot requires a change of materials.

Making something heavy is easy: just leave a hollow space, then fill it with water, sand, or lead shot. Making something light and strong is harder, but often important. Automakers try to make cars lightweight, aircraft manufacturers try harder, and with spacecraft manufacturers it is an obsession. Reducing mass saves materials and energy.

The strongest materials in use today are mostly made of carbon. Kevlar, used in racing sails and bulletproof vests, is made of carbon-rich molecular fibres. Expensive graphite composites, used in tennis rackets and jet aircraft, are made using pure-carbon fibres. Perfect fibres of carbon-both graphite and diamond-would be even better, but can't be made with today's technology. Once molecular manufacturing gets rolling, though, such materials will be commonplace and inexpensive.

What will these materials be like? To picture them, a good place to start is wood. The structure of wood can vary from extremely light and porous, like balsa wood, to denser structures like oak. Wood is made by molecular machinery in plants from carbon-rich polymers, mostly cellulose.

Molecular manufacturing will be able to make materials like these, but with a strength-to-weight ratio about a hundred times that of mediocre steel, and tens of times better than the best steel. Instead of being made of cellulose, these materials will be made of carbon in forms like diamond. Diamond is emphasized here not because it is shiny and expensive, but because it is strong and potentially cheap. Diamond is just carbon with properly arranged atoms. Companies are already learning to make it from natural gas at low pressure. Molecular manufacturing will be able to make complex objects of the stuff, built lighter than balsa wood but stronger than steel.

Products made of such materials could be startling by our present standards. Objects could be made that are identical in size and shape to those we make today, but simultaneously stronger and 90 percent lighter.

This is something to keep in mind next time you're lugging a heavy object around. (If something needs weight to hold it in place, it would be more convenient to add this ballast when the thing is in its proper location than to build in the extra weight permanently.) Better structural materials will make aircraft lighter, stronger, and more efficient, but will have the greatest effect on spacecraft.

Today, spacecraft can barely reach orbit with both a safety margin and a cargo. To get there at all, they have to drop off parts like boosters and tanks along the way, shedding weight.

Fast Development

In some areas of high technology-spaceflight has been a notorious example-it takes years, even decades, to try a new idea. This makes progress slow to a crawl. In other areas-software has been a shining example-new ideas can be tested in minutes or hours. Since the Space Shuttle design was frozen, personal computer software has come into existence and gone through several generations of commercial development, each with many cycles of building and testing.

Fast, Inexpensive Testing

Even in the days of the first operational molecular manipulators, experimentation is likely to be reasonably fast. Individual chemical steps can take seconds or less. Complex molecular objects could be built in a matter of hours. This will let new ideas be put into practice almost as fast as they can be designed.

Later assemblers will be even faster. At a millionth of a second per step, they will approach the speed of computers. And, as nanotechnology matures, experimenters will have more and more molecular instruments available to help them find out whether their devices work or not. Fast construction and fast testing will encourage fast progress.

At this point, the cost of materials and equipment for experiments will be trivial. No one today can afford to build Moon rockets on a hobby budget, but they can afford to build software, and many useful programs have been the result. There is no economic reason why nanomachines couldn't eventually be built with a hobby-size budget, though there are reasons-to be discussed in later chapters-for wanting to place limits on what can be built.

Prior Simplicity

Finally, established technologies are always pushing up against some limit; the easy opportunities have generally been exploited. In many fields, the limits are those of the properties of the materials used and the cost and precision of manufacturing. This is true for computers, for spacecraft, for cars, blenders, and shoes.

For software, the limits are those of computer capacity and of sheer complexity (which is to say, of human intelligence). After molecular manufacturing develops certain basic abilities, a whole set of limits will fall, and a whole range of developments will become possible. Limits set by materials properties, and by the cost and precision of manufacturing, will be pushed way back. Competition, easy opportunities, and fast, low-cost experimentation should combine to yield an explosion of new products.

This does not mean immediately, and it does not apply to all imaginable nanotechnologies. Some technologies are imaginable and clearly feasible, yet dauntingly complex. Still, the above considerations suggest that a wide range of advances could happen at a brisk pace.

The main bottleneck might seem to be a shortage of knowledgeable designers-hardly anyone knows both chemistry and mechanical design-but improving computer simulations will help. These simulations will let engineers tinker with molecular-machinery designs, absorbing knowledge of chemical rules without learning chemistry in the usual sense.

Climbing Complexity

Making familiar products from improved materials will increase their safety, performance, and usefulness. It will also present the simplest engineering task. A greater change, though, will result from unfamiliar products made possible by new manufacturing methods. In talking about unfamiliar products, a hard-to-answer question arises: What will people want?

Products are typically made because their recipients want them. In our discussions here, if we describe something that people won't want, then it probably won't get built, and if it does get built, it will soon disappear. (The exceptions-fraud, coercion, persistent mistakes-are important, but in other contexts.) To anchor our discussion, it makes sense to look not at totally new products, but instead at new features for old products, or new ways to provide old services.

This approach won't cover more than a fraction of what is possible, but will start from something sensible and provide a springboard for the imagination. As usual, we are describing possibilities, not making predictions. The possibilities focused on here arise from more complex applications of molecular manufacturing-nanotechnological products that contain nanomachines when they are finished. Earlier, we discussed strong materials. Now, we discuss some smart materials.

The goal of making materials and objects smart isn't new: researchers are already struggling to build structures that can sense internal and environmental conditions and adapt themselves appropriately. There is even a Journal of Intelligent Material Systems and Structures. By using materials that can adapt their shapes, sometimes hooked up to sensors and computers, engineers are starting to make objects they call "smart." These are the early ancestors of the smart materials that molecular manufacturing will make possible.

Today, we are used to having machines with a few visible moving parts. In cars, the wheels go around, the windshield wipers go back and forth, the antenna may go up and down, the seat belts, mirrors, and steering wheel may be motor-driven. Electric motors are fairly small, fairly inexpensive, and fairly reliable, so they are fairly common. The result is machines that are fairly smart and flexible, in a clumsy, expensive way.

In the Desert Rose scenario, we saw "tents" being assembled from trillions of submicroscopically small parts, including motors, computers, fibres, and struts. To the naked eye, materials made from these parts could seem as smooth and uniform as a piece of plastic, or as richly textured as wood or cloth-it is all a matter of the arrangement and appearance of the submicroscopic parts. These motors and other parts cost less than a trillionth of a dollar apiece. They can be quite reliable, and good design can make systems work smoothly even if 10 percent of a trillion motors burn out. Likewise for motor-controlling computers and the rest. The resulting machines can be very smart and flexible, compared to those of today, and inexpensive, too. When materials can be full of motors and controllers, whole chunks of material can be made flexible and controllable. The applications should be broad.

Surfaces surround us, and human-made surfaces-walls, roofs, and pavement-cover huge areas that matter to people. How can smart materials make a difference here?

The revolution in technology has come and gone, and you want to repaint your walls. Breathing toxic solvents and polluting water by washing brushes have passed into history, because paint has been replaced with smarter stuff. The mid-twentieth century had seen considerable progress in paints, especially the development of liquids that weren't quite liquid-they would spread with a brush, but didn't (stupidly) run and drip under their own weight. This was an improvement, but the new material, "paperpaint," is even more cooperative.

Paperpaint comes in a box with a special trowel and pen. The paperpaint itself is a dry block that feels a lot like a block of wood. Following the instructions, you use the pen to draw a line around the edge of the area you want to paint, putting an X in the middle to show where you want the paint to go on; the line is made of nontoxic disappearing ink, so you can slop it around without staining anything. Using the trowel, you slice off a hunk of paperpaint-which is easy, because it parts like soft butter to the trowel, even though it behaves like a solid to everything else. Very high IQ stuff, that.

Next, you press the hunk against the X and start smoothing it out with the trowel. Each stroke spreads a wide swath of paperpaint, much wider than the trowel, but always staying within the inked line. A few swipes spreads it precisely to the edges, whereupon it smooths out into a uniform layer. Why doesn't it just spread itself? Experience showed that customers didn't mind the effort of making a few swipes and preferred the added control. The paperpaint consists of a huge number of nanomachines with little wheels for rolling over one another and little sticky pads for clinging to surfaces. Each has a simple, stupid computer on board. Each can signal its neigbours.

The whole mass of them clings together like an ordinary solid, but they can slip and slide in a controlled way when signaled. When you smooth the trowel over them, this contact tells them to get moving and spread out. When they hit the line, this tells them to stop.

If they don't hit a line, they go a few handbreadths, then stop anyway until you trowel them again. When they encounter a line on all sides, word gets around, and they jostle around to form a smooth, uniform layer. Any that get scraped off are just so much loose dust, but they stick together quite well.

This paint-stuff doesn't get anything wet, doesn't stain, and clings to surfaces just tightly enough to keep it from peeling off accidentally. If some experimentally minded child starts digging with a stick, makes a tear, and peels some off, it can be smoothed back again and will rejoin as good as new. The child may eat a piece, but careful regulation and testing has ensured that this is no worse than eating plain paper, and safer than eating a colourful Sunday newspaper page.

Many refinements are possible. Swipes and pats of the trowel could make areas thicken or thin, or bridge small holes (no more Spackling!). With sufficiently smart paperpaint, and some way to indicate what it should do, you can have your choice of textures. Any good design will be washable, and a better design would shed dirt automatically using microscopic brushes. Removal, of course, is easy: either you rip and peel (no scraping needed), or find that trowel, set the dial on the handle to "strip," and poke the surface a few times. Either way, you end up with a lump ready to pitch into the recycling bin and the same old wall you started with, bared to sight again.

Looking further at the human environment we find a lot of cloth and related materials, such as carpeting and shoes. The textile industry was at the cutting edge of the first industrial revolution, and the next industrial revolution will have its effects on textiles.

With nanotechnology, even the finest textile fibres could have sensors, computers, and motors in their core at little extra cost. Fabrics could include sensors able to detect light, heat, pressure, moisture, stress, and wear, networks of simple computers to integrate this data, and motors and other nanomechanisms to respond to it. Ordinary, everyday things like fabric and padding could be made responsive to a person's needs-changing shape, colour, texture, fit, and so forth-with the weather and a person's posture or situation. This process could be slow, or it could be fast enough to respond to a gesture. One result would be genuine one-size-fits-all clothing (give or take child sizes), perfectly tailored off the rack, warm in winter, cool and dry in summer; in short, nanotechnology could provide what advertisers have only promised. Even bogus advertising gives a clue to human desires.

Throughout history, the human race has pursued the quest for comfortable shoes. With fully adjustable materials, the seemingly impossible goal of having shoes that both look good and feel good should finally be achieved. Shoes could keep your feet dry, and warm except in the Arctic, cool except in the tropics, and as comfortable as they can be with a person stepping on them.

Adaptive structures will be useful in furniture. Today, we have furniture that adapts to the human body, but it does so in an awkward and incomplete manner. It adapts because people grab cushions and move them around. Or a chair adapts because it is a hinged contraption that grudgingly bends and extends in a few places to suit a small range of preferred positions. Occasionally, one sees furniture that allegedly gives a massage, but in fact only vibrates.

These limitations are consequences of the expense, bulkiness, clumsiness, and unreliability of such things as moving parts, motors, sensors, and computers today. With molecular manufacturing, it will be easy to make furniture from smart materials that can adapt to an individual human body, and to a person's changing position, to consistently give comfortable support. Smart cushions could also do a better job of responding to hints in the form of pats, tugs, and punches. As for massage-a piece of furniture, no matter how advanced, is not the same as a masseuse. Still, a typical massage setting on a smart chair would not mean today's "vibrate medium vigorously," but something closer to "five minutes of shiatsu."

This tour through of the potential of smart matter has shown how we could get walls that look and sound as we wish, clothing, shoes, and furniture of greater comfort, and clean solar power. As one might expect, this just scratches the surface.

If you care to think of further applications, here are some ground rules: Components made by molecular manufacturing can be many tens of times stronger than steel, but materials made by plugging many components together will be weaker. For these, strengths in the range of cotton candy to steel seem achievable. The components will be sensitive to heat, and at high temperatures they will break down or burn. Many materials will be able to survive the temperature of boiling water, but only specialised designs would be oven-safe.

Colour, texture, and (usually) sound should be controllable. Surfaces can be smooth and tightly sealed (this takes some cleverness). Motions can be fairly fast. Power has to come from somewhere; good sources include electricity, stored chemical energy, and light. If nanomachines or smart materials are dunked in liquids, chemical energy can come from dissolved molecules; if they are in the open, energy can come from light; if they are sitting in one place, they can be plugged into a socket; if they are moving around in the dark, they can run on batteries for a while, then run down and quit. Within these limits, much can be accomplished.

"Smart" is a relative term. Unless you want to assume that people learn a lot more about intelligence and programming, it is best to assume that these materials will follow simple rules, like those followed by parts of drawings on computer screens. In these drawings, a picture of a rectangle can be commanded to sprout handles at its corners; pulling a handle stretches or shrinks the rectangle without distorting its right-angle corners. An object made of smart matter could do likewise in the real world: a box could be stretched to a different size, then made rigid again; a door in a smart-material wall could have its position unlocked, its frame moved a pace to the left, and then be returned to normal use. There seems little reason to make bits of smart matter independent, self-replicating, or toxic. With care, smart matter should be safer than what it replaces because it will be better controlled. Spray paint gets all over things and contains noxious solvents; the paperpaint described above doesn't. This will be a characteristic difference, if we exercise our usual vigilance to encourage the production of things that are safe and environmentally sound.

It may be fun to discuss wondrous new products, but they won't make much difference in the world if they are too expensive. Besides, many people today don't have decent food, clothes, and a roof over their heads, to say nothing of fancy "nanostuff."

Costs matter. There is more to life than material goods, but without material goods life is miserable and narrow. If goods are expensive, people strive for them; if goods are abundant, people can turn their attention elsewhere. Some of us like to think that we are above a concern for material goods, but this seems more common in the wealthy countries. Lowering manufacturing costs is a mundane concern, but so are feeding people, housing them, and building sewage systems to keep them from dying of cholera and hepatitis. For all these reasons, finding ways to bring down production costs is a worthy goal. For the poor, for the environment, and for the freeing of human potential, costs matter deeply. Let's take a closer look at the costs of molecular manufacturing.

Inflation produces the illusion that costs rise, when the real story is that the value of money is falling. In the short term, real costs usually don't change very quickly, and this can produce the illusion that costs are stable facts of nature, like the law of gravity or the laws of thermodynamics. In the real world, though, most costs have been falling by a crucial measure: the amount of human labour needed to make things. People can afford more and more, because their labour, supplemented by machines, can produce more and more. This change is dramatic measured on a scale of centuries, and equally dramatic across the gulf between Third World and developed countries. The rise from Third World to First World standards of living has raised income (dropped the cost of labour time) by more than a factor of ten. What can molecular manufacturing do?

Larger cost reductions have happened, most dramatically in computers. The cost of a computer of a given ability has fallen by roughly a factor of 10 every seven years since the 1940s. In total, this is a factor of a million. If automotive technologies had done likewise, a luxury car would now cost less than one cent. (Personal computer systems still cost hundreds of dollars both because they are far more powerful than the giant machines of the 1940s and because the cost of buying any useful computer system includes much more than just the cost of a bare computer chip.)

Real Account

Some costs apply to a kind of product, regardless of how many copies of it are made: these include design costs, technology licensing costs, regulatory approval costs, and the like. Other costs apply to each unit of a product: these include the costs of labour, energy, raw materials, production equipment, production sites, insurance, and waste disposal. The per-kind costs can become very low if production runs are large. If these costs stay high, it will be because people prefer new products for their new benefits, despite the cost-hardly cause for complaint. The more basic and easier to analyze costs are per-unit costs. A picture to keep in mind here is of Desert Rose Industries, where molecular machinery does most of the work, and where products are made from parts that are ultimately made from simple chemical substances. Let's consider some cost components.

***Energy*:** Manufacturing at the molecular scale need not use a lot of energy. Plants build billions of tons of highly patterned material every year using available solar energy. Molecular manufacturing can be efficient, in the sense that the energy needed to build a block of product should be comparable to the energy released in burning an equivalent mass of wood or coal. If this energy were supplied as electricity at today's costs, the energy cost of manufacturing would be something like a dollar per kilogram. We'll return to the cost of energy later.

***Raw Materials*:** Molecular manufacturing won't need exotic materials as inputs. Plain bulk chemicals will suffice, and this means materials no more exotic than the fuels and feedstocks that are, for now, derived from petroleum and biomass-gasoline, methanol, ammonia, and hydrogen. These typically cost tens of cents per kilogram. If bizarre compounds are used, they can be made internally. Rare elements could be avoided, but might be useful in trace amounts. The total quantity of raw materials consumed will be smaller than in conventional manufacturing processes because less will be wasted.

***Capital Equipment and Maintenance*:** As we saw in the Desert Rose scenario, molecular manufacturing can be used to build all of the equipment needed for molecular manufacturing. It seems that this equipment-everything from large vats to submicroscopic special-purpose assemblers-can be reasonably durable, lasting for months or years before being recycled and replaced. If the equipment were to cost dollars per kilogram, and produce many thousands of kilograms of product in its life, the cost of the equipment would add little to the cost of the product.

Waste Disposal: Today's manufacturing waste is dumped into the air, water, and landfills. There need be no such waste with molecular manufacturing. Excess materials of the kind now spewed into the environment could instead be completely recycled internally, or could emerge from the manufacturing process in pure form, ready for use in some other process. In an advanced process, the only wastes would be leftover atoms resulting from a bad mix of raw materials.

Most of these leftover atoms would be ordinary minerals and simple gases like oxygen, the main "waste" from the molecular machinery of plants. Molecular manufacturing produces no new elements-if arsenic comes out, arsenic must have gone in, and the process isn't to blame for its existence. Any intrinsically toxic materials of this sort can at least be put in the safest form we can devise for disposal. One option would be to chemically bond it into a stable mineral and put it back where it came from.

Labour: Once a plant is operating, it should require little human labour (what people do with their time will change, unless factories are kept running as bizarre hobbies). Desert Rose Industries was run by two people, yet was described as producing large quantities of varied goods. The basic molecular-scale operations of manufacturing have to be automated, since they are too small for people to work on. The other operations are fairly simple and can be aided by equipment for handling materials and information.

Space: Even a manufacturing plant based on nanotechnology takes up room. It would, however, be more compact than familiar manufacturing plants, and could be built in some out-of-the-way place with inexpensive land. These costs should be small by today's standards.

Insurance: This cost will depend on the state of the law, but some comparisons can be made. Improved sensors and alarms could be made integral parts of products; these should lower fire and theft premiums. Product liability costs should be reduced by safer, more reliable products. Employee injury rates will be reduced by having less labour input. Still, the legal system in the United States has shown a disturbing tendency to block every new risk, however small, even when this forces people to keep suffering old risks, which are sometimes huge. (The supply of lifesaving vaccines has been threatened in just this way.) When this happens, we kill anonymous people in the name of safety. If this behaviour raises insurance premiums in a perverse way, it could discourage a shift to safer manufacturing technologies. Since such costs can grow or shrink independent of the real world of engineering and human welfare, they are beyond our ability to estimate.

Sales, Distribution, Training: These costs will depend on the product: Is it as common as potatoes, and as simple to use? Or is it rare and complex, so that determining what you need, where to get it, and how to use it are the main problems? These service costs are real but can be distinguished from costs of the thing itself. To summarize, molecular manufacturing should eventually lead to lower costs.

The initial expense of developing the technology and specific products will be substantial, but the cost of production can be low. Energy costs (at present prices) and materials costs (ditto) would be significant, but not enormous. They were quoted on a per-kilogram basis, but nanotechnological products, being made of superior materials, will often weigh only a fraction of what familiar products do. (Ballast, were it needed, will be dirt-cheap.) Equipment costs, land costs, waste-disposal costs, and labour costs can be low by the very nature of the technology.

Costs of design, regulation, and insurance will depend strongly on human tastes and are beyond predicting. Basic products, like clothing and housing, can become inexpensive unless we do something to keep them costly. As the cost of improved safety falls, there will be less reason to accept unsafe products. Molecular manufacturing uses processes as controlled and efficient as the molecular processes in plants. Its products could be as inexpensive as potatoes. This may sound to good to be true (and there are downsides, as we'll discuss), but why shouldn't it be true? Shouldn't we expect large changes to come with the replacement of modern technology?

A Cycle of Falling Cost

The above estimate made a conservative assumption about future costs: that energy and materials will cost then what they do now, before molecular manufacturing has become available. They won't, because lower costs lead to lower costs.

Let's say that making one kilogram of product by molecular manufacturing requires one dollar for a kilogram of raw materials and four dollars for a generous forty kilowatt-hours of energy. These are typical present-day prices for materials and electrical energy. Assume, for the moment, that other costs are small. One of the resulting five-dollar-per-kilogram products can be solar cell paint suitable for applying to paved roads. A layer of paint a few millionths of a metre thick would cost about five cents per square metre to produce, and would generate enough energy to make another square metre of paint in less than a week, even allowing for night-time and moderate cloud cover. The so-called energy payback time would thus be short.

Let's assume that this smart paint costs as much to spread and hook up as it does to make, and that we demand that it pay for itself in a single month, so we charge ten cents per square metre per month. At that rate, the cost of solar energy from resurfaced roads would be roughly \$0.004 per kilowatt hourless than a twentieth the energy cost assumed in the initial production-cost estimate. By itself, this makes the cost of production fall to a fraction of what it was before. Most of that remaining fraction consists of the cost of materials.

But the products of nanotechnology will mostly be made of carbon (if present expectations are any guide), and carbon dioxide is too abundant in the atmosphere these days. With energy so cheap, the atmosphere can be used as source of carbon (and of hydrogen, nitrogen, and oxygen).

The price of carbon would be a few cents per kilogram-roughly a twentieth the original price assumed for raw materials. But now, both energy and raw materials are a twentieth the original price, and so the products become cheaper, including the energy-producing products and the raw-material-producing (atmosphere-cleaning) products.... The above scenario is simple, but it seems realistic in its basic outlines: lower costs can lead to lower costs. How far this process can go is hard to estimate precisely, but it could go far indeed.

Too Cheap Power

This argument will remind some readers of an old claim-that nuclear energy would lead to "power too cheap to metre." This assertion, attributed to the early nuclear era, has passed into folklore as a warning to be skeptical of technologists promising free goodies. Does the warning apply here?

Anyone claiming that something is free doesn't really understand economics. Using something always has a cost equal to the most valuable alternative use for the thing. Choosing one alternative sacrifices another, and that sacrifice is the cost. As economist Phillip K. Salin says, "There's no such thing as a free opportunity," since opportunities always cost (at least) time and attention. Nanotechnology will not mean free goodies.

But, one might argue, nuclear power hasn't even been inexpensive. If technologists could be so wrong back then, why believe a similar argument today? We are happy to report that the arguments aren't similar: any argument for "nuclear power too cheap to metre" had to be absurd even given the knowledge at the time, and our argument isn't.

Nuclear reactors boil water to make steam to turn turbines to turn generators to drive electrical power through power lines to transformers to local power lines to houses, factories, and so forth. The wildest optimist could never have claimed that nuclear power was a free source of anything more than heat, and a realist would have added in the cost of the reactor equipment, fuel, waste disposal, hazards, and the rest.

Even our wild optimist would have had to include the cost of building the boiler, the turbines, the generators, the power lines, and the transformers, and the cost of maintenance on all these. These costs were known to be a major part of the cost of power, so free heat wouldn't have meant free power. Thus, the claim was absurd the day it was made-not merely in hindsight.

In the early 1960s, Alvin Weinberg, head of the Oak Ridge National Laboratory, was a strong advocate of nuclear power, and argued that it would provide "cheap energy." He was optimistic, but did his sums. First, he assumed that nuclear-power plants could be built a little more cheaply than coal-fired power plants of the same size.

Then he assumed that the cost of fuel, waste disposal, operations, and maintenance for nuclear plants would be not much more than the cost of operations and maintenance alone for coal plants. Then he assumed that they might last for more than thirty years.

Finally, he assumed that they would be publicly operated, tax free at low interest (which merely moves costs elsewhere) and that after thirty years the cost of the equipment would be written off (which is an accounting fiction). With all of that, he derived a power cost that "might be" as low as one half the cost of the cheapest coal-fired plant he mentions. He was clearly an optimist, but he didn't come close to arguing for power too cheap to metre.

Nanotechnology–Advancement in Space Technology

The National Space Society believes that developing molecular nanotechnology will advance the exploration and settlement of Space. Present manufacturing capability limits the performance, reliability, and affordability of space systems, but the bottom-up approach of molecular nanotechnology has the potential to produce space hardware with tremendous improvement in performance and reliability at substantially lower cost.

Background

Molecular nanotechnology "expresses the concept of ultimately being able to arrange atoms in a predetermined fashion by manipulating individual atoms" [Aono]. Its principles were first espoused by Nobel prize winner Richard Feynman in 1959, when he said "the principles of physics, as far as I can see, do not speak against the possibility of maneuvering things atom by atom" [Feynman].

As an engineering discipline, molecular nanotechnology promises revolutionary advances not only in manufactured products, but in the processes used to make them. It is the culmination of many fields:

- Microtechnology strives to build smaller devices;
- Chemistry strives to synthesize more complex molecules;
- Molecular biology strives to manipulate with greater precision the wide range of molecular phenomena that occur in living organisms;

- Materials science strives to make stronger, lighter, and more useful solids; and
- Manufacturing strives to build better products for lower cost.

Each of these fields reaches its ultimate in precise, molecular control, which is the ability to build large structures to complex, atomic specifications by direct positional selection of reaction sites. These systems should be able to assemble any configuration of atoms, limited only by the laws of nature and human knowledge — hence they are called universal assemblers.

Because these assemblers would themselves be made of atoms, and because they would be able to assemble these atoms in arbitrary ways, they should be able to self-replicate, or make copies of themselves. NASA, SSI (Space Studies Institute) and others have recognised the potential impact of applying self-replication to space exploration and development [NASA, SSI, Merkle], but have found that self-replication is difficult for macro-molecular devices, partially because each subcomponent level must deal with errors caused at lower sublevels [Neumann, Toth-Fejel, Freitas and Gilbreath].

Space exploration and development has benefited enormously from the advances in these fields, especially microelectronics and materials science, because they reduce payload mass and because they improve reliability. As we converge on the ability to control matter with atomic precision, space development can probably become one of the first and foremost beneficiaries. Molecular nanotechnology can be confused with the micromachines being produced by microlithographic processes, but the two are very different.

Recent Research

The principles of molecular nanotechnology are being demonstrated daily in government and industry laboratories world wide, including the arrangement of 35 xenon atoms to spell out "IBM" [Eigler], the construction of three-dimensional structures from DNA [Seeman], and then engineering of branched, non-biological protein with enzymatic activity [Hahn]. Computer software designed for aiding the development of molecular nanotechnology is also proceeding through the use of tools such as computer-aided design and modelling software [Merkle2].

Potential Benefits and Risks

Near Term Benefits: Since the settlement of Space is not a near-term endevour, it would be a grave mistake to consider only the short term applications of molecular nanotechnology to Space, though there may be a few. In the near term, the chief benefits would most likely be in basic research. For example, improved scanning probes similar to Scanning Tunneling Microscopes (STM) could give researchers a powerful, general technique for characterizing the atomic structure of molecular objects. Such capabilities would be valuable in discovering and designing stronger materials, faster and smaller electronics, and exotic chemicals with unique properties. These incremental improvements would offer the possibility of small improvements in capability across the broad spectrum of space activities, ensuring mission completion, prolonging spacecraft life, and fostering the safety of human crews.

As nanosystems used in research are constructed and commercialized, they will move from gathering basic knowledge in laboratories to collecting data in engineering applications. The first applications would be those in which the relatively high cost and limited capabilities of these first generation devices will still provide significant improvements in overall system capability to justify the costs. Since sensors and actuators could be significantly reduced in size and mass, planetary probes and other space-based applications would probably one of the first beneficiaries of these nanosystems.

Medium Term Benefits

In the medium term, the nanosystem devices would be directly involved in the manufacturing process. Products might include bulk structures such as spacecraft components made of a diamond-titanium composite, or other "wonder" materials. The theoretical strength-to-density ratio of matter is about 75 times that currently achieved by aerospace aluminum alloys, partially because current manufacturing capability allows macro-molecular defects that weaken the material. The bottom-up approach promises to virtually eliminate these defects, enabling the fabrication of stronger materials that could improve reliability and reduce spacecraft dry weight, resulting in increased payload capacity and higher orbital altitude, ultimately reducing the cost to orbit.

In the electronics arena, devices might use a few atoms to store a bit of information (as already demonstrated at IBM). In addition, VLSI (Very Large Scale Integration) would shrink by three magnitudes and extend in three dimensions instead of just two. At this stage, molecular nanotechnology would likely continue to improve capabilities, increase reliability, and lower costs in a wide variety of space projects.

These projected advances would expand the complexity/reliability tradeoff envelope for orbital and lunar systems. Tiny, inexpensive inertial guidance systems could assist unmanned exploratory spacecraft, planetary rovers, and interplanetary probes. A dense network of distributed embedded sensors throughout a manned or unmanned spacecraft could continuously monitor (and affect, if they could be operated as actuators) mechanical stresses, temperature gradients, incident radiation, and other parameters to ensure mission safety and optimize system control.

In an advanced spacecraft, the outer skin would not only keep out the cold and the vacuum, but it might also function as a multi-sensor camera and antenna. With such extensive monitoring and increasingly efficient control of propulsion systems, life support, and other spacecraft systems, mission success rates would increase at lowered cost. Advanced materials may also enhance on-orbit human activities by providing more effective spacesuits, and may foster more extraterrestrial endeavours by developing more efficient and degradation-resistant solar cells.

As capabilities increase, the molecular techniques used in the actual manufacturing of the spacecraft (i.e. the tools and processes that transform raw materials into advanced sensors and materials) would themselves increase in capability. This advance would make it much easier to build spacecraft systems that could take advantage of in-situ extraterrestrial resources.

Long Term Benefits

Since the settlement of Space is a long term enterprise, these long-term benefits of molecular nanotechnology are the most relevant. And these benefits are considerable. The most important arises from the general ability to build nanosystems, especially the ability to bootstrap production via self-replicating universal assemblers.

This capability would probably lower manufacturing costs by many magnitudes, down to the order of $1 per kilogram. It would also make possible to build tapered tethers from geosynchronous orbit to the ground, and to build human-rated SSTO vehicles with a dry mass around sixty kilograms.

Such capabilities should make possible inexpensive access to space. Mature nanosystems might make possible affordable and robust closed environment life-support systems that could take advantage of in-situ resources, such as asteroidal metals and cometary organics.

Such a capability would potentially enable many people to affordably live in space. Tiny computers, sensors and actuators, trivially cheap on a per-unit basis, may allow things like smart walls to automatically repair micrometeorite damage, comfortable and unobtrusive space suits, and terraforming tools. By providing instrumentation that allows the development of medical knowledge at the molecular level, advanced nanosystems might enable in vivo repair of cellular damage.

This capability should mitigate the dangers of ionizing cosmic radiation.

Further long term effects of this technology are completely unpredictable, but would undoubtedly be quite significant. Absolute and relative costs will still constrain space activities, however, and some desired activities will remain impossible.

Costs and Risks

Before applications can be developed for the exploration and development of Space, molecular nanotechnology itself must become an practical discipline instead of just a theoretical one. There are three promising paths to the building of universal assemblers: genetic engineering, physical chemistry, and scanning probe microscopy. Uncertainty remains as to which path is easiest and quickest, and hybrid approaches appear quite promising, so efforts should be spread across these three areas.

There are likely to be a very large number of expensive blind alleys, so it is important to not invest too much money in any one area. Japan is aggressively pursuing molecular nanotechnology by investing approximately $200 million over 10 years with industry matching government funding in over twenty companies, while funding for the Atomcraft Project [Aono] is being continued by six Japanese companies as it completes its five year plan.

Attempts are underway to start a similar program in Switzerland. If the U.S. fails to engage in activities leading to expanded and well-implemented research with commercially relevant goals, we will probably find ourselves critically behind in the broader economic, military, and specific space-related benefits that may accrue from these technologies. The cost of trailing behind in this technology would be very high.

Many potential threats consist of someone using molecular nanotechnology for aggressive purposes. Thus, efforts must be undertaken to ensure that both global security and U.S. national security are safe against this potential threat. One strategy for ensuring US and global security is to develop molecular nanotechnology in a collaborative, multi-lateral manner. This addresses fears of many nations that they will be caught behind in development, and maintains trust since open collaboration is de facto open and mutual inspection.

There is a possibility that due to some unforeseen law of science, universal assemblers may be impossible to build. In this case, the risk consists of a zero return on investment. But in the ten years since the concepts of molecular nanotechnology have been made public, no one has proposed any scientific reasons for its impossibility. The absence of these reasons might be explained by the fact that numerous objects around us (all carbon-based life-forms) have been formed using the bottom up approach, and by the fact that long-range trends in technology show a continually increase in the precision with which matter can be controlled.

Space versus Molecular Manufacturing

There is a fear that spending money on molecular nanotechnology will reduce the amount of money spent on Space development, since research funding is sometimes perceived as a zero sum game. One version of this argument asks why the small amount of money available for research should be spent on speculative ventures such as molecular nanotechnology when projects such as DC-X seem to be much closer to success.

- First, decision theory and experience show that achieving large projects of significant technological complexity (e.g. the settlement of Space) require a diversification of effort. It is especially important to have a diversified portfolio of approaches so that unforeseen dead ends can be circumvented without delay. In this case, Space development can benefit significantly by investing a limited amount of effort in low cost, high risk, and high payoff avenues such as molecular nanotechnology.
- Second, the amount of money needed at this stage of molecular nanotechnology development is very small compared to the average NASA Space project.
- Third, much of basic science research, especially the biological and material sciences, concerns itself with the behaviour of increasing small groups of atoms. In addition to having nothing in common with Space development, almost all of this research focuses on science, not engineering. Since this research is occuring anyway (regardless of the NASA's and DOD's budget, and hence not part of a zero sum game) it seems that a small effort could steer it in directions most likely to create a space-faring civilization.
- Fourth, the settlement of Space, like the development of molecular nanotechnology, is a long term enterprise, so care must be taken not to sacrifice the long term goals of the latter to achieve short term goals of the former, especially when advanced nanosystems could significantly increase our capabilities in Space.

The opposite version of the "space versus molecular manufacturing" argument ask why money should be spent on expensive Space hardware when exploring and developing Space would be less costly with advanced molecular nanotechnology. While it is true that nanosystems could significantly lower the cost of Space missions, other factors must be considered.

- First and most important, policy makers will undoubtedly make decisions in the near future about molecular nanotechnology, and these decisions should be informed by an acceptance of permanent human presence and expansion into the high frontier of Space. If these decisions are instead made with the assumption that humanity is limited to Earth, the results will most probably be catastrophic.
- Second, it is not known how quickly nanosystems will reach maturity, nor how much effort will be directed toward including them in the design of Space applications. Therefore it seems prudent to continue

Space activities and utilise nanotechnologies as they come on line.

- Third, the absence of a significant human direction toward Space may allow social inertia (including cultural attitudes toward frontiers, civil and criminal law, and levels of technical education) to become a major obstacle in developing nanosystems for Space applications.
- Fourth, Space provides a frontier in which advances in molecular manufacturing will mature with many social impacts. If those social impacts occur in the context of a closed society, it is likely that the Western enlightenment values of humanism, reason, and science will die.[Zubrin]

Conclusion

In conclusion, the National Space Society believes that since the settlement of Space is a long range project that will benefit the entire human race, the serious development of the long range field of molecular nanotechnology must be supported. Extraterrestrial activities are a natural application for nanosystems, and synergistic effects between Space and Molecular Nanotechnology can and should be encouraged.

7

Nanotechnology and Nanotechnology Development

Nanotechnology is a powerful new technology for taking apart and reconstructing nature at the atomic and molecular level. It is being touted as the basis of the next industrial revolution and will be used to transform and construct a wide range of new materials, devices, technological systems and even living organisms. Nanotechnology will likely underpin and impact all industries and sectors of the economy, and is likely to facilitate far-reaching changes in social, economic and ecological relations. Opinion is sharply divided regarding whether these changes will be largely positive or negative.

Proponents suggest that nanotechnology will deliver gains in fields as diverse as manufacturing, medicine, environmental remediation and military applications. However critics argue that nanotechnology introduces serious new risks to human health and the environment, raises problematic ethical issues and is likely to result in large-scale socio-economic disruption. Governments are beginning to recognise the need for new laws to protect workers, the public and the environment from the risks of nanotoxicity.

However despite the commercial availability of over 720 products containing nanomaterials, not a single government world-wide has yet introduced regulations that require nanomaterials to be subject to new safety assessments prior to commercial release.

The failure of government regulators to take seriously the early warning signs surrounding nanotoxicity suggests that they have learnt nothing from any of the long list of disasters that resulted from the failure to respond to early warning signs about previous perceived "wonder" materials (like asbestos, DDT and PCBs).

Nanotechnology is being commercialised largely outside of general public awareness or debate, and without any serious attempt to involve the community in decision making about its introduction. Issues of ethics, democracy and nanotechnology's broader socio-economic impacts have yet to register in the debate.

What is Nanotechnology?

There is still not an internationally accepted nomenclature, set of definitions and measurement systems for nanotechnology, although work towards these has begun. The lack of a standardised nomenclature and measurement system has made it difficult to compare safety tests, exposure measurement and risk assessment carried out to date. However, the term 'nanotechnology' is now generally understood to encompass both nanoscience and the broad range of technologies that operate at the nanoscale.

- *Nanoscience:* The study of phenomena and materials at the atomic, molecular and macromolecular scales, where properties differ significantly from those at the larger scale
- *Nanotechnology:* Design, characterisation, production and application of structures, devices and systems by controlling shape and size at the nanoscale
- *Nanoscale:* Having one or more dimensions of the order of 100nm or less, or having at least one dimension that affects functional behaviour at this scale
- *Nanomaterials:* Particles, nanotubes, nanowires, quantum dots, fullerenes (buckyballs) etc that exist at a scale of 100nm or less, or that have at least one dimension that affects their functional behaviour at this scale

One nanometre (nm) is one thousandth of a micrometre (m), one millionth of a millimetre (mm) and one billionth of a metre (m).

To put 100 nanometres in context: a strand of DNA is 2.5nm wide, a protein molecule is 5nm, a virus particle 150nm, a red blood cell 7,000 nm and a human hair is 80, 000 nm wide and a flea is around 1,000,000nm in size.

Engineered vs. Incidental Nanoparticles

Engineered nanoparticles are deliberately manufactured and can be distinguished from nanoparticles that 'exist in nature', or are by-products of other human activities. 'Incidental' nanoparticles (also called ultrafine particles in the study of air pollution and its epidemiology) are a by-product of forest fires and volcanoes, and high-temperature industrial processes including combustion, welding, grinding and vehicle combustion. It is the manufactured or engineered nanotechnological products and processes that are the primary focus of the issues raised in this briefing paper. However many of the safety and regulatory issues relating to manufactured nanoparticles are also relevant to incidentally produced nanoparticles (e.g. the need to establish safe workplace exposure limits for all types of nanoparticles).

Nanotechniques

A number of forms of nano-techniques can be distinguished, including:

- *Nanoparticle Production:* This includes production of carbon nanotubes, "buckyballs", quantum dots, and the nano-scale manufacture of previously larger scale materials and chemical compounds (including metal oxides such as titanium, zinc, aluminium, manganese, iron, silver, gold and more)
- *Nanofabrication and Molecular Manufacturing:* these include a range of existing and hypothetical techniques and processes for assembling systems and structures from the atom up. The goal of molecular manufacturing is to create independent small-scale nanofactories that will enable efficient, decentralised industrial manufacture with atomic level precision
- *Nanobiotechnology:* This includes the use of nanotechnology to manipulate living organisms and to enable the merging of biological and non-biological materials Why is nanotechnology attracting so much interest?

What is Novel about Nanotechnology?

The fundamental properties of matter change at the nanoscale. The properties of atoms and molecules are not governed by the same physical laws as larger objects or even larger particles, but by "quantum mechanics". The physical and chemical properties of nanoparticles can therefore be quite different from those of larger particles of the same substance. Altered properties can include but are not limited to colour, solubility, material strength, electrical conductivity, magnetic behaviour, mobility (within the environment and within the human body), chemical reactivity and biological activity.

The altered properties of nano-sized particles have created new possibilities for profitable products and applications. The use of nanoparticles, the potential of nanofabrication and molecular manufacturing, and new breakthroughs associated with synthetic biology and the melding of living and non-living organisms, has generated a lot of excitement in the research and business communities. Oxford University Professor George Smith has been quoted as joking that "nano is from the Greek verb meaning 'to attract research funding'"

Why are People Touting Nano as the Next Industrial Revolution?

Nanotechnology embodies the dream of controlling the building blocks of both living and non- living things, and the ability to remake the world from the atom up. The excitement around nano is building because people believe that it will bring changes as significant or potentially more far reaching than those that accompanied the European Industrial revolution. Proponents and critics alike suggest that nanotechnology will enable break throughs in a wide number of different fields- communications, agriculture, cognitive science, medicines, military and environmental remediation to name a few...

"Convergence of diverse technologies is based on material unity at the nanoscale and on technology integration from that scale... [Nanoscience and nanotechnology] will allow us to understand and, when desirable, to control the behaviour both of complex microsystems, such as neurons and computer components, and macrosystems, such as human metabolism and transportation vehicles"

> *"The real power of nanoscale science is the potential to converge disparate technologies that can operate at this scale. With applications spanning all industry sectors, technological convergence at the nanoscale is poised to become the strategic platform for global control of manufacturing, food, agriculture and health in the immediate years ahead"*

Why should we be sceptical about nanotechnology's ability to solve our problems?

Nanotechnology proponents suggest that it may provide a panacea for growing problems of climate chaos, water shortages, pollution, poverty, disease and social unrest. Nano optimists see nanotechnology providing: universal clean water supplies; greater productivity in agriculture and nutritionally enhanced foods; cheap and powerful solar energy generation; clean and highly efficient manufacturing; and radically improved formulation of drugs, delivery modes, diagnostics and effective vaccines for most diseases. However industrial-technological solutions alone cannot fix problems stemming from socio-economic inequity or the unequal distribution of power. Nanotechnology, like other new echnologies, does not exist in isolation from its economic or political context. Nano investment and applications to date remain driven by the less altruistic economic and political realities of the 'real world'.

The first nanoproducts to be released commercially are targeted squarely at wealthy consumers in the global north, including: anti-wrinkle cosmetics; odour-eating socks; superior display screens for computers, televisions and mobile phones; premium coatings for luxury cars; and self-cleaning windows and bathrooms.

The US government, which is the world's biggest funder of nano research, directs the largest portion of its funding to military applications. In 2006, the US defence program received a third of the total US$1.3billion investment in nanotechnology, which was a greater share than that received by the National Science Foundation

Our experience of the European Industrial revolution, and other periods of rapid industrial and technological change, tells us that the benefits of nanotechnology will not be shared equally. Very little attention has been paid to studies of the likely disruptive impacts for the world's poorest people if and when nanomaterials displace existing commodity markets (eg for rubber, cotton or copper) and cause massive job loss

Now, as during the previous industrial revolution, it will be the world's poorest people who are least able to adapt quickly in the face of technological change.

Overview of Investment and Commercial Value

Investment in nanotechnology is growing rapidly. Combined investment in nanotechnology by the government and private sectors is growing rapidly from year to year. About US$10 billion was spent on nanotechnology research and development in 2004, which is almost double the money spent in 2003. At least 60 countries have established national nanotechnology research programmes, about half of which are in Europe.

In 2004, Japan led the way, investing almost US$4 billion, with the US not far behind at US$3.4 billion. The combined investment from Germany, the UK, France and Italy was about half this. Following that in order of descending investment, was: Taiwan, South Korea, China and Australia. In Australia, about US$200 million was invested, with about equal proportions coming from government and the private sector.

The growth of private sector investment in nanotechnology is now outpacing that of government. 2003 was the first year that private sector funding about equalled public funding, but in 2004 the corporate sector's investment was more like two-thirds of the total money invested. Nanotechnology cuts across all industry sectors and investment reflects this. Virtually all the Fortune 500 companies now run nano-programmes. To put the scale of nanotech investment in context, the US government's US$5 billion for nano research and development between 2001 and 2006 makes it the biggest publicly funded science endeavour since the Apollo moon landing.

Nanotechnology is already making a commercial impact In 2004, global sales of nanotechnology products totalled US$10 billion. The overwhelming majority of this revenue came from sales of nanoscale materials, with a very small amount coming from nanodevices. Industry analysts Lux Research estimated that products incorporating nanomaterials constituted around 0.1% of global manufacturing output.

Future predictions of industry growth are necessarily speculative and vary widely. In 2001, the US National Science Foundation estimated that by 2015 the global nano industry would be worth US$1trillion, employ 2 million workers directly, and contribute to the production of half the world's new manufactured products. However growth figures continue to be revised upwards, with some recent comments indicating the NSF believes the US$1trillion figure may be reached as early as 2011.

In 2004 industry analysts Lux Research estimated that by 2014 products incorporating nanomaterials would constitute around 15% of global manufactured goods, with a total value of as much as US$2.6trillion. To put this into context, that would approach the size of the information technology and communication industries combined. Nanoproducts- what's available now and what can we expect in the future Products containing nanomaterials already on the market

Although some people still describe this phase of nano's commercialisation as being "pre-competitive", more than 720 products containing nanomaterials are already available.

Products containing nanomaterials have been released commercially in the absence of regulatory oversight. There is no legal requirement anywhere in the world for manufacturers to conduct new safety tests on nano-scale ingredients to ensure that the people who use these products and the workers involved in product manufacture, packaging and transport, are not exposed to unacceptable risks. There is similarly no requirement for manufacturers to demonstrate that the novel properties of nano-scale ingredients, nano-processing, or product manufacture do not present an increased negative impact to the environment. There is no requirement for manufacturers to indicate the inclusion of nanoscale ingredients on product labels.

The following list represents just a fraction of the products that are already on the market:

- Transparent sunscreens
- Cosmetics including lipsticks, face powders, hair conditioners, moisturizers and anti-ageing creams
- Temperature moderating, stain, moisture and odour-repellent clothing
- Food additives
- Food packaging

- Agricultural fertilisers
- Long-lasting paints furniture varnishes and car coatings
- Self-cleaning windows and building surfaces
- Computer chips and mobile phones
- Inks
- Magnetic recording tapes and memory storage devices
- Optical fibres
- Chemical-mechanical polishing
- Land-mine detectors
- Solid-state compasses
- Fuel cells
- Industrial catalysts
- Specialist automotive and aerospace components
- Broad range of military applications
- Display technology for laptops, mobile phones, digital cameras
- Football stadium lights
- Metal-cutting tools
- Self-cleaning surfaces for glass and building surfaces
- Glare-reducing coatings for eyeglasses and car windscreens
- Automatic catalysts converters
- Bumper bars and step assists for cars
- Tennis balls and racquets
- Dental-binding agents
- Burn and wound dressings
- Bio-imaging products
- Nanomaterials for environmental remediation
- Disinfectants and anti-bacterial products
- Anti-graffiti coatings for walls

What will Nanotechnology Look Like in the Future?

It is difficult to imagine exactly how nanotechnology development is likely to unfold and what its timeline may be. Even the industry experts revise their forecasts every few years. We know that the use and production of nanomaterials will expand significantly over the coming years, with implications for human health, the environment and broader trade relations. However it is hard to know how many of the hypothetical higher-tech applications of nanotechnology will actually come to fruition, and when. Some of the current and hypothetical applications of nanotechnology are described briefly below.

Present day: Applications based primarily on the use of passive and active nanomaterials for their novel properties. This draws on well-established branches of applied science including materials science. Key applications are in coatings, pigments, electronics and photonics and biotechnology.

Medium-term (2015): Half of all newly designed advanced materials and manufacturing processes may be built using control at the nanoscale, with increased used of nanoscale devices. Creation of three-dimensional nanosystems and the development of precise molecular assembly. Healthcare and life sciences applications become significant as nano-enabled pharmaceuticals and medical devices emerge from lengthy human trials.

Technology convergence enables vastly superior treatment of disease (including an effective treatment for cancer) and life extension, including via the production of synthetic organs. 'Smart' foods interact with consumers to 'personalise' food, changing colour, flavour or nutrients on demand or in response to an individual's allergies or nutrient needs.

Longer term (2015-2050): Development of molecular assembly-based nanofactories capable of decentralised, atomically-precise manufacture of everything from bicycles to supercomputers to weapons. Atomic-level genetic control of crops, plants and animals; use of ubiquitous nano surveillance and monitoring systems on farms increases productivity and reduces labour needs.

Fast, broadband interfaces directly between the human brain and machines that transform work in factories, control automobiles, ensure military superiority to its early developers, and enable new sports, art forms and modes of interaction between people.

"Instead of harvesting grain and cattle for carbohydrates and protein, nanobots could assemble the desired steak or flour from carbon, hydrogen and oxygen atoms present in the air as water and carbon dioxide"

Why the Science of the Small Brings Huge Problems?

The expectations and hype surrounding nanotechnology have fuelled a nano gold rush. Governments and corporations world-wide have scrambled to be part of the 'next industrial revolution', not wanting to be left on the wrong side of a future divide between 'nano-haves' and 'nano-have nots'.

In the midst of this race to boost research, seek patents and commercialise as quickly as possible, serious questions regarding ethics, human and environmental safety, socio-economic disruption and democracy have been ignored. Some key areas where unresolved questions remain include:

Serious Ethical Problems

Ethical problems underlie nanotechnology's quest to manipulate the very building blocks of life; its aggressive commercialization enabled by research carried out with public monies, but driven by commercial and military interests; and the failure of governments to halt the rapid introduction of nanoproducts and nanomaterials until serious public interest issues are addressed adequately. Nanobiotechnology raises significant ethical concerns in its quest to engineer organisms and manufactured products containing both biological and human-made components.

The US National Science Foundation's work to use convergent nanotechnology, biotechnology, information technology and cognitive science to improve human performance beyond species-typical boundaries is also particularly ethically problematic. This work has drawn strong criticism from disabilities and human rights advocates concerned that it will create new inequities and further marginalise existing disadvantaged groups.

Risks to Human and Environmental Safety

There is a growing body of toxicological evidence that nanoparticles present serious new risks to human and environmental health. Leading scientific organisations, including the United Kingdom's Royal Society, have warned that the risks of nanotoxicity are serious. In 2004 the Royal Society recommended that nanomaterials should be treated as new chemicals and be subject to new safety assessments prior to their inclusion in consumer products. The Royal

Society further recommended that factories and research laboratories should treat nanomaterials as if they were hazardous, and until the environmental impacts of nanomaterials are better known, their release into the environment should be avoided as far as possible. And yet no government world-wide has introduced a regulatory system to protect the health of workers, the public and the environment from the risks associated with nanotoxicity.

The Royal Society clearly recommended prohibiting the deliberate release of nanomaterials for bioremediation until its ecological implications were better understood, and yet this is already taking place. Concerns surrounding the potential for deliberate or unintentional release of self-replicating organisms that could cause ecological damage cannot be ruled out.

Socio-Economic Disruption

Very little attention has been paid to studies of the likely disruptive impacts, and massive job losses, associated with the expansion of the nanotechnology industry and its displacement of existing industries. Novel nanomaterials could replace markets for existing commodities, disrupt trade and eliminate jobs in nearly every industry. Industry analysts Lux Research

Inc. have warned that nanotechnology will result in large-scale disruption to commodity markets and to all supply and value chains: "Just as the British industrial revolution knocked handspinners and handweavers out of business, nanotechnology will disrupt a slew of multi billion dollar companies and industries". This would have the most devastating impact on people in the developing world whose countries are dependent on trade in raw resources (e.g. rubber, cotton, copper) that were displaced by nanomaterials. There are clearly profound impacts for labour associated with the promise of molecular manufacturing systems. In the nearer term, the introduction of nanosensors, along with increasingly automated production, has the potential to drive down demand for on-farm and manufacturing labour.

A Deadly Nano Arms Race

Nanotechnology will provide the tools for ubiquitous surveillance, with significant implications for civil liberties. The growing nano arms race may enable a whole new generation of weapons of mass destruction including nano-biological weaponry. The expansion of nanoweaponry is truly alarming, especially given the enormous difficulties associated with establishing reliable safeguar ds against its use. Retired Admiral David Jeremiah of the US navy believes nanotechnology will prove more significant than nuclear weapons in determining future political power relations. Military research and development is already attracting the lion's share of nanotechnology funding from the US government, which is the world's largest single investor in nanotechnology. In the 2006 US$1.3billion budget for the

US National Nanotechnology Initiative, the US Department of Defence received $436 million (33.5% of the nanotechnology budget). Conversely, only $38.5million (less than 4%) was earmarked for both the study of the health, safety and environmental impacts of nanotechnology, and also potential applications in these areas.

Erosion of Democracy

Despite the huge transformative potential of nanotechnology, and the billions of dollars of public funding invested in research and development, there is little transparency in how the decisions shaping this technology are being made. There has been no effort to engage the public in decision making about its introduction or regarding how the billions of dollars of public monies should be invested. Friends of the Earth believes that it is unacceptable that the introduction of such a transformative new technology should be driven solely by business and political interests, with so many outstanding questions and without the involvement of the broader community.

Friends of the Earth Australia is calling for an immediate moratorium on all commercial research, development, production and release of nanotechnological materials and products. Given the serious risks and impacts associated with nanotechnology's introduction, public involvement in decision making regarding nanotechnology and the introduction of a regulatory regime based on the precautionary principle must be prerequisites to further commercialisation of nanoproducts.

It is essential that civil society has an informed debate about whether or not it actually wants the changes that nanotechnology will bring, and has the opportunity to be involved in decision making about public policy and regulatory development. To manage nanotechnology's risks, we recommend the development of a comprehensive regulatory regime in which:

- All nano-materials and products are subjected to rigorous health and environmental impact assessment, including evidence based testing, prior to commercial production and/ or environmental release
- Due to the radically altered characteristics of nano materials compared to their larger scale counterparts, nano materials are assessed as new substances, even where the properties of larger scale counterparts are well-known
- The assessments are based on the precautionary principle and the onus is on proponents to prove safety, rather than relying on an assumption of safety
- Risk assessment includes the entire life cycle of the products in question, from 'cradle to grave'
- All relevant data related to safety assessments, and the methodologies used to obtain them, are placed in the public domain
- Nanotechnological products are subjected to a social impact assessment to ensure the development, application and control of nanotechnologies do not reinforce or create new forms of socio-economic inequalities, concentrations of wealth and power, means of social control and oppression, or weapons of destruction
- Nanotechnological products are subject to an ethical assessment within a robust framework developed with widespread public input
- Public participation in decision making on nanotechnology and other new technologies

We recognise that further scientific research on the health and environmental safety of nano-materials and products is required to inform the development of regulations to manage the risks of nanotoxicity. However, for even this non-commercial research work to proceed, precautionary protocols will be required to protect the health and safety of researchers and to strictly minimise any environmental releases.

Nanotechnology and its Future Applications

This is a technology involving the potential ability to fabricate structures and devices with atomic precision by controlling the size of the matter at the scale of 1-10nm.

It will provide the solution to a large number of problems faced by mankind today. A nanometre is one billionth of a metre (10-9), roughly the width of three or four atoms.

The Potential of Nanotechnology

The potential of nanotechnology is huge and can lead to tremendous miniaturization in wider areas like space systems, medial diagnostic equipments and drug delivery systems. It will enable us to fabricate very sensitive devices and machines, leading to the enhancement of human capabilities to work efficiently, at lowers cost, with more precision and in environmentally friendly ways.

Nanotechnology will make it feasible for us to create such sophisticated devices and structures with more flexibility at nanoscale.

Areas in which nanotechnology has future applications and discoveries, which can lead to enormous economical and industrial development, is as follows:

- Macromolecular design and folding
- Self-assembly methods
- Catalysis (inorganic, enzyme and other)
- Dendrimers, fullerenes and other novel chemical structures
- Bioenergetics, nanobatteries and ultrasound-driven chemistry
- Semiconductor-organic/biological interfaces
- Miniaturization and massive parallelism of SFM
- Molecular modelling tool

Applications:

- Energy Storage, Production and Conversion:
  a) Novel hydrogen storage systems based on carbon nanotubes and other lightweight nanomaterials
  b) Photovoltaic cells and organic light-emitting devices based on quantum dots
  c) Carbon nanotubes in composite 0.lm coatings for solar cells
  d) Nanocatalysts for hydrogen generation
  e) Hybrid protein-polymer biomimetic membranes

- Agricultural Productivity Enrichment:
  a) Nanoporous zeolites for slow release and efficient dosage of water and fertilizers for plants and of nutrients and drugs for livestock
  b) Nanocapsules for herbicide delivery
  c) Nanosensors for soil quality and for plant health monitoring
  d) Nanomagnets for removal of soil contaminants
- Water Treatment and Remediation:
  a) Nanomembranes for water purification, desalination and detoxification
  b) Nanosensors for the detection of contaminants and pathogens
  c) Nanoporous zeolites, nanoporous polymers and attapulgite clays for water purification
  d) Magnetic nanoparticles for water treatment and remediation
  e) TiO2 nanoparticles for the catalytic degradation of water pollutants
- Disease Diagnosis and Screening:
  a) Nanoliter systems (Lab-on-a-chip)
  b) Nanosensor arrays based on carbon nanotubes
  c) Quantum dots for disease diagnosis
  d) Magnetic nanoparticles as nanosensors
  e) Antibody-dendrimer conjugates for diagnosis of HIV-1 and cancer
  f) Nanowire and nanobelt nanosensors for disease diagnosis
  g) Nanoparticles as medical image enhancers
- Drug Delivery Systems:
  a) Nanocapsules, liposomes, dendrimers, buckyballs, nanobiomagnets and attapulgite clays for slow and sustained drug release systems
- Food Processing and Storage:
  a) Nanocomposites for plastic.lm coatings used in food packaging
  b) Antimicrobial nanoemulsions for applications used in decontamination of food equipment or packaging
  c) Nanotechnology-based antigen detecting biosensors for identification of pathogen contamination
- Air Pollution and Remediation:
  a) TiO2 nanoparticle-based photocatalytic degradation of air pollutants in self-cleaning systems
  b) Nanocatalysts for more efficient, cheaper and better-controlled catalytic converters

   c) Nanosensors for detection of toxic materials and leaks
   d) Gas separation nanodevices

- Construction- nanomolecular structures to make asphalt and concrete more robust to counter water seepage:
   a) Heat-resistant nanomaterials to block ultraviolet and infrared radiation
   b) Nanomaterials for cheaper and durable housing, surfaces, coatings, glues, concrete and heat and light exclusion
   c) Self-cleaning surfaces (e.g. windows, mirrors, toilets) with bioactive coatings
- Health monitoring Nanotubes and nanoparticles for glucose, CO(2), and cholesterol sensors and for in-site monitoring of homeostasis:
- Vector and pest detection and control:
   a) Nanosensors for pest detection.
   b) Nanoparticles for new pesticides, insecticides and insect repellents

Nanotechnology for Site Remediation

The chapter contains information attained from a wide variety of currently available sources, including project documents, reports, periodicals, Internet websites, and personal communication with both academically and commercially employed sources. No attempts were made to independently confirm the resources used. It has been reproduced to help provide federal agencies, states, consulting engineering firms, private industries, and technology developers with information on the current status of this project.

According to an encyclopedic source, nanotechnology is defined as the design, characterization, production, and application of structures, devices, and systems by controlling shape and size at the nanoscale. Nanotechnology is an emerging science that already shows promise in improving various life aspects ranging from medicine to industrial materials.

Along with our ambition to move forward with nanotechnology, we as a society must be conscious of the potential environmental and health impacts of nanomaterials. It naturally results that our best interest is in studying the inherent impact of nanomaterials on us as a society, an ecosystem, and an integral part of nature. This document focuses on the use of nanomaterials in site remediation. As is well known to researchers and scientists of the field, nanotechnology has great potential in performing chemical and physical processes that are useful in toxic chemical remediation.

It is observed that otherwise recalcitrant species found in the environment are neutralized at unprecedented rates and efficiencies when treated with specialised nanoparticles such as nanoscale zero-valent iron (NZVI). With nanotechnology being as cost effective as it is relative to other leading methods, it becomes necessary to understand its potential impact on the environment. Nine current EPA Office of Research and Development (ORD) Science to Achieve Results (STAR) grants aim at increasing our understanding of the fate and transport of nanomaterials as they are used for desirable processes in the environment. The immediate question of concern becomes: do we understand the physiochemical properties of nanoscale materials well enough to effectively apply them towards remediation? This document will attempt to answer this question by providing information on recent research.

Background

Nanoscale materials have been tested at a number of contaminated sites with preliminary reports of success. The sites mentioned herein contain chlorinated hydrocarbons such as TCE (trichloroethylene), PCE (perchloroethylene), and other recalcitrant compounds often found as contaminants of concern at EPA's National Priorities List sites. The original concept of using metals as a remediation agent is based on a widely known chemical reaction that uses a neutral electron donor (a metal) to chemically reduce an electron acceptor (a contaminant). The overall reaction is called a reduction-oxidation reaction, or simply a "redox" reaction. In its inception, redox-based remediation viewed elemental iron as the ideal electron donor because of its abundance and ease of application and preparation. Scientists immediately became aware of the potential for metals as reducing agents and sought more effective methods of preparation and application.

Where the original treatment was macroscopic and solely iron-based, new treatments utilise "manufactured" iron in which remediation takes place on the nanoscale. One specific method utilises advanced iron particles coated with noble metals (colloidal bimetallic nanoparticles, or BNP) that act as catalysts in the redox reaction. In many cases, other elements with properties similar to those of iron are used depending upon site-specific remediation requirements. It has been reported that even zinc or silver can be used as an electron donor in the metal-oxide remediation scheme. Ideal properties of a reductive material are that it be metallic (neutral or "zero valent") and that it holds a reductive potential of two standard charges. While the chemical properties of a redox-based process are vital to remediation, the physical properties are equally important and play an even more important role in a nanoscale environment. Zero-valent iron remediation traditionally is applied via a permeable reactive barrier (PRB). This technology involves placement of reactive media in one or more soil trenches located directly in the path of contaminated ground-water flow, thereby providing direct contact between contaminants and the reducing agent. Depending upon solubility and other characteristics of a target contaminant, as well as site-specific geochemical conditions, redox reactions can occur successfully within the PRB.

An alternate ZVI-based remediation technology employs reactive media in the form of slurry. The slurry is injected directly into the subsurface through one or more injection wells strategically placed within the target treatment zone. This approach helps mobilize aqueous-phase iron and increase distribution of reactive material across the treatment zone. Both methods have advantages; while PRBs are particularly effective where treating a well-defined contaminant plume, injection methods are useful in treating source areas or a relatively immobile contaminant plume.

Case Studies of NZVI-Based Remediation

Nease Chemical, OH

Based on pilot-scale and bench-scale NZVI studies, decision-makers at many sites across EPA's 10 regions are considering use of this technology to remediate contaminated ground water and soil. One example is the Nease Chemical Site in Ohio, where NZVI technology is one of five cleanup options under consideration by EPA Region 5. The Nease site is geographically sectioned into areas, or "operable units," needing specific treatment. The property consists of relatively flat terrain with unlined ponds that serve as a source of chemical seeps. The treatment areas are known as "ponds 1 and 2," "remaining ponds and soil," and "shallow and deep ground water." A June 2005 proposed remediation plan and site description specifies NZVI technology employing palladium-coated colloidal iron. The reactive medium will be administered through injection wells reaching deep ground water containing a plume of volatile organic compounds (VOCs).

RCRA Site, AK

Another site where NZVI is being considered for remediation purposes is an Alaskan facility regulated under EPA's Resource Conservation and Recovery Act (RCRA) program. The site contaminants include VOC constituents of diesel fuel, which was released during an earlier tanker spill, in addition to PCE and TCE. Research indicates that ZVI may effectively reduce the chlorinated compounds in ground-water plumes as well as soil. Field-scale tests will be conducted in both shallow and deep treatment zones. If field-scale tests yield the anticipated positive results, the project will be expanded to full-scale operation.

Manufacturing Plant, Trenton, NJ

The site of a manufacturing plant in Trenton, NJ, was selected for testing of emulsified zero-valent iron (EZVI) technology to address a contaminant plume that had migrated downgradient from the facility. The test employed

ZVI in the form of nanoscale particles suspended in a hydrophobic fluid, thus providing micelles to act as a means of transporting NZVI to non-aqueous phase liquid (NAPL). The micelle membranes are bound by a surfactant that allows cell transport through the ground-water plume.

EZVI was injected in a water-based slurry through an injection well ("DGC-15") also serving as a monitoring well. Three multi-depth monitoring wells, or piezometres ("PZ-1," "PZ-2," and "PZ-3") also were installed. After the injection process, researchers noted significant TCE reductions in ground water of the second monitoring well but little concentration change in the third.

According to an associated Environmental Science & Technology journal article, the primary purpose of this field study was to obtain NZVI "proof of concept," i.e., evidence that ZVI administered in the slurry was nanoscale in dimension. The following redox reaction was provided by the authors to show how BNP are manufactured on the nanoscale through the process of reductive deposition:

$$Fe_0 + Pd_2 + \rightarrow Fe_2 + + Pd_0$$

Reduction of a chlorinated compound can then proceed via the following reaction:

$$C_2HCl_3 + 4Fe_0 + 5H+ \rightarrow C_2H_6 + 4Fe_2 + + 3Cl_{-4}$$

In some cases, up to 96% reductions in TCE concentrations were observed. Other studies show similar results from ZVI-based treatment, and significant reductions in typical contaminant plumes.

Klockner Road Site, Hamilton Township, NJ

This site also is known as the "PSE&G Site" due to its proximity to the PSE&G Trenton (NJ) Switchyard. The area of contamination covers approximately four acres of uncultivated and vacant property historically used as a disposal area for various industrial wastes. Geologically, the site exhibits a perched water aquifer 2-8 feet below ground surface (bgs) and a bedrock formation 130-160 feet bgs. Administration of ZVI reactive agent is performed through a 20-point injection system on a square grid. Monitoring is conducted on the same grid system to gain a better understanding of chemical reduction that is anticipated to occur via the following reactions:

$$3Fe_0 + 3Fe_2 + + 6e- \rightarrow CCl_3CH_3 + 6e- + 3H+ \; C_2H_6 + 3Cl\text{-}Fe_0 + 2H_2O \rightarrow Fe_2 + + H_2 + 2OH\text{-}CCl_3CH_3 + 3H_2 \rightarrow C_2H_6 + 3HCl$$

NASA Launch Complex 34, Cape Canaveral, FL

This site is well known for its historical use as a launch pad for shuttle craft and other space-bound vehicles using rocket propulsion. Typical with the use of rocket fuel, chlorinated compounds such as TCE exist in the site's ground water, soil, and sediment.

Field studies suggest that bioremediation via natural attenuation will enhance the results of active treatment technologies implemented at this site.

Three methods of administering NZVI will be field-tested at Launch Complex 34 as alternatives to the direct-injection method: pressure pulsing, pneumatic fracturing, and hydraulic fracturing. Pressure pulsing involves forced administration of NZVI particles to various and relatively predictable subsurface depths. Pneumatic fracturing employs compressed air to create subsurface crevices and small pathways that facilitate distribution of the reactive medium. In a way similar to pneumatic fracturing, hydraulic fracturing uses high-pressure liquids to enhance reagent distribution in the subsurface.

Each of these methods is anticipated to show reductions in ground-water contaminant plumes.

Focus on Nanoparticle Fate and Transport

Issues: Although funding and individuals conducting research in this area exist, nanoscale technology is novel and extensive knowledge about nanoparticle fate and transport does not exist. The question regarding "where the particles go" after serving their remediation purpose is significant, particularly as vendors increasingly produce nanoscale metal products and site managers increasingly show interest in using nanotechnology for site remediation. An alternate approach to answering this question is to pose a related question: how can researchers track these nanosized particles in a macroscopic environment? To approach a solution, material scientists are trying to understand the behaviour of nanoscale materials in sedimentary environments.

Research

Many ORD-grant research teams studying nanotechnology fate and transport focus their efforts on soil systems. Some researchers focus on nanotechnology applications in the environment (as nanoparticles occur naturally), while yet others work to increase the understanding of nanoparticle bioaccumulation. Additionally, many research teams have shown that nanoparticles may aggregate or agglomerate to form larger macro-scale metal particles with better known impacts on the environment. These agglomerated nanoparticles may pose concern regarding uptake in wildlife and bioaccumulation. In this sense, the prominent concern becomes the physical interactions between the particles themselves and the environment.

Case Studies of Fate and Transport of Nanoparticles for Site Remediation

Manufacturing Plant, Trenton, NJ

This manufacturing plant synthesizes bimetallic particles through use of a process based on the following chemical reaction:

$$4Fe_3{++}\ 3BH_4 + 9H_2O \rightarrow 4Fe_0 + 3H_2BO_3 + 12H{+} + 6H_2$$

In order to achieve the bimetallic catalytic effect, iron and palladium are combined in a weight ratio of 1:300. The reactive catalyst coating is applied via the following reaction: Fe0+ Pd2+ ® Fe2++ Pd0 BNP material is introduced to the contaminated ground-water plume via injection wells, and the reactive agent is distributed throughout the target area via simple suspension in ground water. Monitoring at the piezometres suggests a significant discrepancy in migration rates of the plumes after BNP injection. (Piezometres use pressure differences between at least two subsurface levels to generate a correlation of relative water levels. This method relies on the fact that different plumes exert different pressures although their linear velocities may be the same.) Data provided by PARS Environmental suggest that the contaminant plume continued to move at a rate of 0.3 m/day while the nanoparticles were observed to travel at a rate of 0.8 m/day. Although the relative rates show discrepancy, the impact of the difference is negligible when considering the benign nature of iron particles in soil and water systems.

Klockner Road Site, Hamilton Township, NJ

A patented BNP-based product known as NanoFe Plus™, developed by PARS Environmental, Inc., is used for remediation at the Klockner Road Site.

Environmental, Inc., reports that the reactive agent is administered in a water-based slurry containing one pound of reactive material in each 3.994 gallons of solution.

Currently, little information is available on the fate and transport of NanoFe Plus particles in the subsurface environment at this site. The injection phases and quantities of the injections are known, and two monitoring sessions are scheduled to take place after two injection phases. Monitoring will be conducted to determine the amount of dense nonaqueous-phase liquid reduced by the nanoscale iron, and to estimate the quantity of iron remaining in the aquifer. According to PARS Environmental, little evidence exists showing that the iron plume has travelled farther than the contaminant plume.

A slightly different application method is used for EZVI technology targeting chlorinated solvents at Launch Complex 34. This method involves the mixing of nanoscale reactive particles with edible oil to form an emulsion that enhances particle mobility within the contaminant plume. EZVI nanoparticles are made using a 1:55 (weight) ratio of palladium and iron. NASA Kennedy Space Centre reports that measures of oxidation-reduction potential indicate rates of EZVI migration are negligible.

Conclusions

Preliminary results in the case studies indicate that NZVI-based technology has potential as a means of site remediation. When accompanied by sound environmental safeguarding, the technology may prove to be safe and reliable.

Currently, more research is needed regarding potential impacts of nanoparticles and their fate and transport in environmental media. According to researchers and developers, nanoparticles will become more complex in their design and use. This raises the same question regarding the extent to which nanoparticles will be environmentally benign when developed through use of synthetic materials such as polymers.

What is known is that the use of complex nanoparticles in remediation is still in the research phase. Remediation technology developers need to recognise that the release of synthesized materials with complex multi-molecular nanoscale structures will generate questions concerning potential ecological and environmental threats.

Based on the information available, the current methods of nanoscale site remediation do not appear to pose a threat to humans and the environment. Transport data indicate that nanoscale particle plumes may travel slightly faster than the natural migration rate of contaminant plumes. Many case studies found the migration rate of iron plumes is insignificant when compared to the rate of contaminant reduction.

Questions to be Answered

This section highlights questions arising during this research. The greatest priority at this stage in nanotechnology is concern about the advent of complex nanoparticles and the extent of their propagation in (and potential impacts to) the environment. To date, there have been only bench-scale studies using nanoparticles formulated from the study of polymer synthesis. The questions to be answered are:

- What impacts, if any, would polymer-based nanoparticles have on the environment and human health?
- What time frame should be considered for full-scale production of polymer nanoparticles?
- What is the potential benefit of using complex nanoparticles versus species such as nanoscale iron or zinc?

Further Research Needs

This section highlights the need for study on other aspects of current nanoscale applications in hopes of inspiring new grants and grantees for research. Specific research is needed to:

- Assess the extent to which nanotechnology implications reach soil and water systems beyond immediate application areas,
- Evaluate the potential impact of nanoparticles on specific and immediate environmental media,

- Determine whether, and the degree to which, nanoparticles may travel through the food chain; preliminary research already indicates that some nanoparticles may be taken up by bacteria in trace amounts, and
- Examine the uptake of nanoparticles in animal species.

Emerging Nanotechnologies for Site Remediation and Wastewater Treatment

The impacts of nanotechnology are increasingly evident in all areas of science and technology, including the field of environmental studies and treatment. Experts anticipate the development and implementation of environmentally beneficial nanotechnologies in the categories of sensing and detecting, pollution prevention, and treatment and remediation. Of the three, the category of treatment and remediation has seemingly experienced the most growth in recent years. In terms of site remediation, the development and deployment of nanotechnology for contaminant destruction has already taken place.

Nanoscale iron particles and the subsequent derivatives (bimetallic iron particles and emulsified iron) represent a viable commercially available nanotechnology for remediation. Currently, over 15 academic and commercial field scale tests involving nano-iron particles are underway or have reached completion. Many more sites have scheduled field studies and consequently the number of vendors supplying this product continues to grow. In addition, a multitude of nanotechnology applications for site remediation and wastewater treatment are currently in the research and development stages. From dendritic polymers to functionalized ceramics, the technologies poised to impact the treatment field are considerably diverse.

For the next few decades, at the very least, our country will be faced with serious issues regarding the cleanup of contaminated sites across the country. Despite a significant effort over the past 25 years, the job remains far from complete and as of yet, many sites remain inadequately characterized. A number of contaminated areas await remedial action, and many still await identification. Adhering to the current regulatory standards, an estimated 294,000 sites (range 235,000-355,000) require some form of remediation; this does not include sites with completed or ongoing remediation projects. Contaminated sites can be divided into seven groups based around regulatory and decontamination responsibility: Superfund, Resource Conservation and Recovery Act (RCRA) Corrective Action, Underground Storage Tanks (USTs), Department of Defence (DOD), Department of Energy (DOE), Civilian Federal Agencies, and States (USEPA 2004).

The majority of these sites, which are both large and complex, will require the collaboration of multiple stakeholders for successful cleanup as well as the development and implementation of innovative remedial solutions. The drive for novel remediation processes is in high demand. Expedited and efficient processes often reduce costs, thus saving the responsible parties a great deal of money. Successful remediation also allows for the ability to reuse land.

Another driver for innovation originated from the 1986 reauthorization of the Superfund law, which added a new emphasis to "permanence and treatment," (USEPA 2004). While the law only encompasses the National Priority List (NPL)/Superfund sites that comprise a relatively small fraction of the total list of contaminated sites, the Superfund program has greatly influenced the entire field of remediation. The idea of permanence and treatment has largely been adopted by all site divisions. In addition, many new and better remediation technologies implemented at Superfund sites have spread to other sectors due to the interchangeability of innovative technologies across the various segments. The remedial compatibility across site divisions, in turn, has created an extremely large market and, thus, a substantial demand for these new technologies (USEPA 2004).

In the past ten years, emerging technologies such as phytoremediation, bioremediation, and permeable reactive barriers have become popular new tools. These novel treatments have begun to compete with more established technologies such as solidification/ stabilization, soil vapor extraction, and thermal desorption for soil, and pump and treat systems for groundwater (USEPA 2004). At the very forefront of these emerging technologies lies the development of nanotechnology for site remediation. Nanotechnology represents an extremely broad field, which encompasses a number of materials and technologies spanning multiple disciplines. Currently a wide variety of potential remedial tools employing nanotechnology are being examined at the bench-scale for use in waste water and soil remediation. One emerging nanotechnology, nanosized zero valent iron and its derivatives, has reached the commercial market for field-scale remediation and studies.

Purpose

While nanotechnology is considered the new buzzword by many in the scientific community, information regarding the subject remains largely dispersed and fragmented due to the relative novelty of the technology. This fact holds true for the specific developments in the field of hazardous waste remediation. While one specific nanotechnology and its derivatives have reached the commercialization process, a majority of the current studies involving nanotechnologies for remedial applications remain on the bench scale. Recent journals and books, such as the Environmental Science and Technology's Special Issue on Nanotechnology and the ACS Symposium Series, Nanotechnology and the Environment, provide a more comprehensive view of the recent advances.

However both of these publications present the majority of information as a compilation of peer-reviewed articles. These types of publications often detail the science behind the technology, leaving the overall relevance and connections to field scale applications vague. In general, a lack of assembled information exists on how these individual nanotechnologies might eventually be implemented, the type of sites where they would be appropriate, and the general feasibility for scale-up. In evaluating the status of this emerging field while also looking ahead to the future, it becomes important to examine both the science behind these nanotechnologies as well as their current and potential applications. This paper seeks to provide a more holistic view of the state of the science. Both the commercialized nanotechnology products and many of the technologies being researched in academia are discussed. Attention is given both to the research itself as well as the remedial capabilities. The toxicity and safety concerns of the individual technologies are also briefly outlined as are the overall toxicity concerns related more generally to the field of nanotechnology. Finally the current state of regulation is addressed.

Within the category of sensing and detection, nanotechnology anticipates the capability to provide more sensitive and cost effective technologies for detecting pollution in the ground, water, and air. Smaller particle size enables the development of smaller sensors, which can be deployed more easily into remote locations (Masciangioli 2003). The high surface area to mass ratio, characteristic of nanoparticles, amplifies changes in electrical conductivity and mechanical resonance that occur when pollutants bind to the particle surface. This feature translates into sensing systems able to detect very small amounts of pollutants (Rose-Pehrsson 2004). Array sensors, capable of detecting various pollutants discriminately, also benefit from nanotechnology. Their ability to manipulate the surface chemistry of particles allows a new level of selectivity control (Rose-Pehrsson 2004). Single-wall nanotubes (SWNTs) are exemplary of the potential developments in this field. The unique chemical and electronic features of SWNTs enable them to act as sensors of electrical resistance changes in the presence of a targeted pollutant, such as nitrogen dioxide. Zinc oxide also is being researched as a potential dual-function sensor and remediator (Kamat 2002).

The ability of nanotechnology to abate pollution production is just beginning to be explored and could potentially catalyze the most revolutionary changes in the environmental field (Masciangioli 2003, Karn 2004). From the birth of manufacturing onward, production has been achieved through a top-down approach. In many ways, top-down production parallels stone sculpture, where an artist begins with a huge block of stone, chiseling, grinding, and sanding to reach a finished product. Not only does this approach produce waste, but it requires large energy expenditures (Karn 2004).

The "1.7 kg microchip" has recently been adopted as the new spokesmodel for the scientific community when discussing the problems of top-down manufacturing (Masciangioili 2003). A study published in 2002 by Williams et al. determined that the production of a single 2-gram, 32-megabyte chip requires the use of 32 kg of water and 1.7 kg of fossil fuels and chemicals (Williams 2002 and Masciangioli 2003). Nanotechnology claims the ability to revolutionize manufacturing through a bottom-up assembly approach, where products self-assemble from molecular building blocks similar to the biological assembly of proteins (Karn 2004). The idea of molecular manufacturing, however, largely remains theoretical and most likely will for the coming years. Many nanotechnologies on the market today do exhibit a level of self-assembly, just not on the same scale required in manufacturing.

With any new technology, however, come concerns over its misuse and possible negative effects. The great excitement for nanotechnology and its potential applications already has been met by apprehensions over environmental and health effects induced by the unintentional release of nanoparticles into the environment. Because nanotechnology produces substances with such unique properties and functionalities, worker exposure and accidental release pose serious potential threats. Similarly, because of the broadness of the field, regulation of nanoparticles will be difficult. At the moment it is unclear whether current regulations are adequate for these emerging technologies.

Commercial Uses of Nanotechnology for Site Remediation

Over the years, the field of remediation has grown and evolved, continually developing and adopting new technologies in attempts to improve the remediation process. One of the most established systems is that termed "pump-and-treat" (USEPA 1998b). Pump-and-treat systems operate on the basis of removing contaminated groundwater from the ground, downstream of the contamination site, and then treating it before returning it to the ground. With this technology "it takes a long time to achieve cleanup goals, it has been demonstrated to spread contamination in some cases, and it is expensive to operate and maintain," requiring continual energy input (USEPA 2004, 1998b).

"A 2001 EPA study found that the average annual operations and management (O&M) costs of pump-and-treat systems at 79 fund-financed sites are approximately $570,000, and the median is $350,000... The periods of operation of these systems as well as the costs vary widely from site to site. The average pump-and-treat system in the EPA study operated for 18 years, for an average cost of $10 million. Pump-and-treat systems at some sites with dense non-aqueous phase liquids (DNAPLs) may need to operate for considerably longer periods." (USEPA 2004). Despite the number of limitations exhibited by these systems, pump-and-treat remedies still account for 67% of the ground water remedies proposed or in progress at NPL sites (USEPA 2004, 1998b).

In the early 1990s, the reducing capabilities of metallic substances, such as zero-valent iron (ZVI), began to be examined for their ability to treat a wide range of contaminants in hazardous waste/water (Zhang 2003). The most common deployment of ZVI has been in the form of permeable reactive barriers (PRBs) designed to intercept plumes in the subsurface and subsequently remediate them (USEPA 1998b). PRBs, first installed at the field-scale in 1994, offer a substitute for the more established pump-and-treat systems (USEPA 1998b). [The first full-scale commercial PRB was approved for use in the State of California by the San Francisco Regional Water Quality Control Board (RWQCB) in 1994. USEPA1998b].

This passive treatment system has been used to treat pollutants, including chlorinated hydrocarbons, nitro aromatics, polychlorinated biphenyls (PCBS), pesticides and even chromate. The reducing capabilities of ZVI can dechlorinate chemicals such as trichloroethene (TCE) and polychlorinated biphenyls (PCBs) (Lien 2001). ZVI can reduce hexavalent chromium (Cr(VI)) to trivalent chromium (Cr(III)), as well, precipitating it out of solution, thus immobilizing it as Cr (III) hydroxides or chromium-iron hydroxide solid solutions (USEPA 1997, 1998b).

While PRBs can potentially replace pump-and-treat systems as better remedial solutions, the technology itself is not without drawbacks. As compared to pump-and-treat systems, installation costs for PRBs are relatively high and the effective "time to replacement" is often uncertain (USEPA 2004). The installation capabilities only allow PRBs to be inserted up to about 50 ft, which restricts the technology to shallow plumes only. To prevent contaminants from circumventing the PRB, plumes must be carefully characterized and delineated (USEPA 1997). Problems occur from a decrease in iron reactivity caused by the precipitation of metal hydroxides and metal carbonates onto the surface of the iron (Wang 1997).

The low reactivity of ZVI toward lightly chlorinated compounds allows the formation and perseverance of hazardous byproducts. Compounds such as cis-1,2-dichloroethane (cDCE) and vinyl chloride often can be observed as products from the reduction of tetrachloroethene (PCE) and trichloroethane (TCE) (Wang 1997). Just as PRBs were designed to provide a better alternative to pump-and-treat, new technologies are now available to compete with PRBs. As detailed below, nanoscale iron particles and their derivatives offer a potentially more effective and economical alternative to many remedial technologies (Zhang 2003). The small particle size of the nano iron (1-100 nm) facilitates a high level of remedial versatility. This allows a much greater diversity in applications as compared to the traditional ZVI employed in PRBs. The elevated surface area to mass ratio, a common characteristic of nanoparticles, also enhances the reactivity of the iron, making it a promising emerging technology.

Technology Overview

Nanoscale Zero Valent Iron (nZVI) and Reactive Nanoscale Iron Product (RNIP) comprise the most basic form of the nano iron technology (Zhang 2003, Okinaka 2004). Particles of nZVI, typically about 100 to 200 nm in diameter, consist solely of zero valent iron (Fe0). The most common route to nZVI synthesis employs sodium borohydride as the key reductant (Zhang 2003). In 1997, Wang et al. first produced the nanoscale iron particles in the laboratory using the method of sodium borohydride reduction. By mixing sodium borohydride (NaBH4) with FeCl3·6H2O, Fe3+ is reduced according to the reaction scheme below:

$Fe(H_2O)_{63+}$+ $3BH_4$ + $3H_2O$ Æ Fe0 + $3B(OH)_3$ + 10.5H2 (Wang 1997) In laboratory scale production of nZVI, Wang et al. achieved a particle size distribution of less than 100 nm for 90% of the particles produced. The BET surface area for the particles was determined to be 33.5 m2/g (Wang 1997). Following the reaction, the reduced particles of iron (Fe0) created could be directly used for contaminant destruction. The stoichiometry of the reduction of trichloroethene (TCE) to ethane, a typical decontamination reaction, would proceed as follow:

C_2HCl_3 + $4Fe_0$ + 5H+ Æ C_2H_6+ $4Fe_{2+}$+ 3Cl-(Elliott 2001) RNIP particles vary slightly from nZVI particles, in that RNIP particles consist of approximately a 50/50 wt% mixture of iron and magnetite (Fe3O4). The core of the particles consists of the elemental iron (a-Fe), while the Fe3O4 surrounds the Fe, forming an outer shell (Okinaka 2004).

In a study by Okinaka et al., particle sizes for RNIP were averaged at about 70 nm and mean BET surface areas were determined to be 28.8 m2/g (Okinaka 2004). The synthesis method for RNIP particles has been patented and detailed, with ferrous sulfate listed as the starting material. The use of ferrous sulfate as a raw material contributes to a small (5,000 mg/kg) sulfur content present in the RNIP particles.

A recent study by Liu et al. compared the efficiency and degradation capabilities of nZVI synthesized using sodium borohydride reduction and the RNIP particles produced from ferrous sulfate. The study found that nZVI demonstrated a higher initial surface area and Fe0 content. It was concluded, though, that the presence of boron and the shell thickness were the most likely explanations for observed differences in reactivity. The nZVI particles demonstrated rapid dechlorination of TCE and no deactivation; however rapid H2 evolution was observed. This behaviour could cause the particles to "burn out" before they could reach the treatment zone. Conversely, RNIP particles showed much slower degradation rates of TCE and a lower accessibility of Fe0 due to the Fe3O4 shell. The shell, however, also protected the particles from reacting with water, thus preventing the "burn out" observed with the nZVI particles (Liu 2005).

Other methods of producing nanosized iron particles also have been developed. Ball milling represents another technique. In this process, micron-size iron powder is reduced to the nanoscale through an attrition or abrasion process using a ball mill (Liles 2004). A vacuum/gas condensation process also has been used to produce nanosized iron and other metals (Canano Technologies 2005). The production of Bimetallic Nonsocial Iron Particles (BNIP) represents an enhancement of the nZVI technology described above. Grittini et al. reported in 1995 that a bimetallic system of palladium and iron could rapidly degrade PCBs (Grittini 1995). While Grittini used microscale iron particles in his experiments, the discovery of palladium's ability to enhance reductive capabilities proved significant. In 1997, Wang et al. created the first nZVI particles. In the same experiment, they also coated the nZVI particles with palladium. Later Zhang et al. determined that the nanoscale palladium-coated ZVI degraded chlorinated compounds at rates 10 to 100 times faster than microscale particles.

To synthesize the particles, after obtaining Fe0 as described above, Wang et al. saturated the iron with an ethanol solution of [Pd(C2H3)2]3 leading to the deposition of palladium on the iron surface through the reaction:

Pd_{2+} + Fe_0 Æ Pd0 + Fe_2 + (Wang 1997) Initial experiments performed by Wang et al. determined that while nZVI particles dechlorinated TCE and PCB compounds at a higher rate than iron powders, palladium-coated BNIP proved the most reactive and thus the destructive to the chlorinated organic contaminants (COCs) (Wang 1997). Scientists have experimented using other noble metal catalysts, such as plutonium, gold, and nickel; however palladium/ iron BNIPs are the only particles commercially available.

As with the addition of metal catalysts to nZVI particles, the formation of emulsified zero valent iron (EZVI) also represents an enhancement to the existing nZVI technology. Emulsion droplets 7 can be created using a food grade surfactant, biodegradable vegetable oil, water, and nZVI, where a surfactant-stabilized oil-liquid membrane forms around the iron particles in water (Quinn 2005). Specifically, in a study performed by Quinn et al., Sunlight-brand corn oil, sorbitan triolate (a nonionic surfactant) and RNIP particles from Toda were used to make EZVI. Both nano and microscale iron can be used; however the studies and information described in recent studies involve only nZVI. Emulsions typically have a diameter of about 40 um and a specific gravity of approximately 1.1 (Quinn 2005, O'Hara 2004).

Remedial Applications

As mentioned previously, the small particle size and high surface area to mass ratio make iron nanoparticles highly reactive and extremely versatile.

The high surface area and surface reactivity compared with granular forms enable the nanoparticles to remediate more material at a higher rate and with a lower generation of hazardous byproducts (Zhang 2003). The ability of the nanoparticles to act as strong reducers also enables the remediation of an extremely wide range of contaminants. Based on the documented case studies detailed below, the majority of field applications have utilised nano iron to remediate chlorinated organics compounds, such as TCE. A growing number of sites, though, contain metal ions such as chromium (Zhang 2003).

Table: *Contaminants remediated by nanoscale iron (Zhang 2003)*

Carbon Tetrachloride	***Chrysoidine***	***Cis-Dichloroethene***
Chloroform	Tropaeolin	trans-Dichloroethene
Dichloromethane	Acid Orange	1,1-Dichloroethene
Chloromethane	Acid Red	Vinyl Chloride
Hexachlorobenzene	Mercury	PCBs
Pentachlorobenzene	Nickel	Dioxins
Tetrachlorobenzenes	Silver	Pentachlorophenol
Trichlorobenzenes	Cadmium	NDMA
Dichlorobenzenes	Bromoform	TNT
Chlorobenzene	Dibromochloromethane	Dichromate
DDT	Dichlorobromomethane	Arsenic
Lindane	Tetrachloroethene	Perchlorate
Orange II	Trichloroethene	Nitrate

In conjunction with nano iron's diverse group of target contaminants, the field scale deployment of the particles can be achieved in a variety of ways. Nanoparticles can be mixed with water to form a slurry that can be injected using pressure or gravity into a contaminated plume (Zhang 2003). Once injected, the particles remain in suspension, forming a treatment zone. Particles of iron also can be used in ex situ slurry reactors to treat soil, sediment, and solid waste. In cases of water and/or wastewater treatment, anchoring nanoparticles onto a solid matrix, such as activated carbon, can prove extremely effective (Zhang 2003). The injection of nano iron into the ground represents the most common deployment of this technology thus far.

Overall the process provides a number of remedial benefits. In comparison with PRBs, nanoparticle injection allows remediation at greater depths and in areas unreachable by PRBs (i.e. land covered by a building). Most importantly, this technique facilitates source zone remediation, a clear benefit for site cleanup. The hydrophilic surface of the nano iron particles only permits the remediation of aqueous phases, which excludes DNAPL remediation.

However nano iron injections into the source still provide benefits unachievable by downgradient technologies. Remediation at the point of dissolution eliminates the migration distance required for PRBs and pump-and-treat systems.

It also has been speculated that the presence of nZVI in the source zone amplifies the concentration gradient between the aqueous phase and DNAPL, which in turn increases the mass transfer of contaminants from DNAPL to the dissolved phase.

To address the significant number of sites across the country contaminated with chlorinated VOCs in the form of DNAPL, researchers developed EZVI. As mentioned previously, NIP injections only work for aqueous phase source zone treatment and while they can enhance DNAPL dissolution, NIP injections cannot directly treat DNAPL. EZVI offers a solution to this problem in the form of an external oil membrane (Quinn 2005, O'Hara 2004).

The EZVI, exhibiting a specific gravity of 1.1, can be considered in some ways a DNAPL itself. Similarities between the exterior membrane and DNAPL allow the EZVI to be miscible with the DNAPL (O'Hara 2004). As a result of this miscibility, when EZVI droplets come in contact with pure phase contaminants, the DNAPL dissolves through the exterior oil membrane. Once the DNAPL passes through the oil membrane, the contaminants become trapped. The confinement of the DNAPL to the emulsion facilitates the degradation by the NIPs. This capability thus can potentially allow source zone remediation of chlorinated DNAPLs (Quinn 2005).

Vendors

In the past five years, a variety of vendors for nZVI and RNIP have appeared. Polyflon Company manufactures a nZVI product called Polymetallix, which is then distributed by Nanitech LLC. PARS Environmental, Inc., manufactures and distributes NanoFe. Wei-xiang Zhang of Lehigh University supplies nZVI particles to many field sites although he does not manufacture his product on a commercial scale. OnMaterials represents the most recent market edition with their product Z-loy. Toda Kogyo Corporation manufactures RNIP, and is the sole producers of this patented technology.

Polymetallix, NanoFe, and Zhang's particles are all prepared in the same manner, using sodium borohydride to reduce Fe3+ to nZVI (Zhang 2003, Bilvey Conversation 2005, PARS 2005). While Zhang achieved particle sizes of less than 100 nm in the lab, for large-scale projects, all three distributors advertise particle sizes of between 100 and 200 nm. The approximate surface area of the nZVI particles is advertised as 37.0-58.0 m^2/g by Polymetallix manufacturers. However, Zhang et al. determined the average surface area of their particles to be approximately 33 m^2/g (Zhang 2003).

OnMaterials employs the second technology outlined above for the manufacture of their product Z-loy. The company website describes the mean particle size as 250 nm and the median particle size as 200 nm, with particle surface areas ranging from 1-30 m2/g (OnMaterials 2005). As described above, Toda's RNIP particles represent a mixture of Fe and Fe304, with an average size of 70 nm and an average surface area of 28.8 m^2/g (Okinaka 2004). Prices for nZVI and RNIP vary between vendors. Nanitech LLC has quoted the cost of Polymetallix at $77 per pound for orders of 300 lbs, and $72 per pound for orders of 400 lbs including delivery. PARS prices orders exceeding 1,000 pounds at $45 per pound for polymer-supported particles and $31 per pound for unsupported material. OnMaterials quotes pilot scale quantities of the Zloy product at $20 per pound with a shipping charge of $3 per pound. Toda estimates the cost of RNIP from $26 to $34 per pound based on the order amount, where the range is 10,000 lbs to 100 lbs respectively (Gavaskar 2005).

Many of the same vendors that manufacture pure nZVI also offer palladium-coated BNIP as well. PARS Environmental, Inc., manufactures NanoFe Plus. OnMaterials advertises the ability to customize orders of Zloy through choosing the ratio of iron to palladium. Wei-xiang Zang also produces BNIP in his lab in addition to nZVI, with the majority of field tests and treatments using BNIP.

In terms of cost, coating with a catalyst often adds significant expense to an order. PARS quotes polymer supported nZVI at $45 a pound, while an order of palladium-coated (and supported) BNIP costs $66 per pound. The near 50% cost increase accompanying palladium coating must be thus factored into the cost-benefit (Gavaskar 2005). Currently no vendors exist for EZVI. Case studies and experiments involving the technology have used EZVI prepared on site. As previously noted, nZVI and RNIP particles act as bases when forming EZVI.

Case Studies

The application of nano iron and its derivatives for site remediation has quickly gained popularity as an option in the field. A number of projects involving some form of nano-iron have been reported, and it appears that these sites represent a variety of the regulatory categories.

The following case studies attempt to provide an overview of the field applications.

Edison New Jersey Industrial Site: The Edison New Jersey location is the site of a retired adhesives manufacturing plant, which operated from 1967 to 2002. Until 1990, the manufacturing process employed trichloroethane (TCA). No major spills or releases were recorded during the operational period. However, contamination has been attributed to small spills incurred during routine transfer of TCA from tanker rail cars to aboveground storage tanks.

A number of chlorinated volatile organic contaminants have been detected in the groundwater at the site in high concentrations, primarily TCA and TCE. Degradation products including DCA, DCE, CA, and vinyl chloride also have been identified (Chu 2005).

The contamination area fell under the classification of fractured bedrock, specifically Brunswick Shale. A thin layer (4-6 ft) of soil comprised mainly of silt and clay covered the bedrock. Both a primary source area and secondary release area were identified. Concentrations as high as 37,000 mg/L of TCA indicated the possible presence of DNAPL in the primary source area. A pump and treat system was installed in July 2001 and was operational up to the point of the study (Chu 2005).

The Secondary Source Area served as the location for the pilot study. The contamination in the area was considered relatively localized, and the area was located away from the buildings and utilities on site. Z-loy, produced by OnMaterials, and emulsified vegetable oil were used concurrently to provide an abiotic/biotic remedy. Concentrated iron product and vegetable oil were mixed together on site; a total of 300 pounds of nZVI and 1,500 gallons of emulsified vegetable oil comprised the mixture. Injection took place in two locations. One injection well was converted from a groundwater extraction well, and the other injection point was converted from a shallow bedrock monitoring well.

The solution was injected into both wells in approximately equal amounts and at pressures between 25 and 50 pounds psi (Chu 2005). Following the injections, both injection well and two downgradient monitoring wells were monitored for 13 months. Results indicated that the nZVI produced rapid abiotic degradation in the injection wells while the vegetable oil encouraged a more lasting biological process downstream in the monitoring wells.

This was inferred from data collected over the monitoring period. Injection Well 1 experienced a sharp drop in TCA concentration from 10,000 ug/L to a level below the minimum detection limit. DCA concentrations experienced an initial increase due to the degradation of TCA, however levels quickly began to decline. CA, the product of DCA degradation, increased for a longer period of time before finally beginning to decrease.

DCE concentrations also decreased with no accumulation of VC or ethene, indicating complete degradation to ethane. Overall ethane concentrations for Injection Well 1 steadily increased. Also of note, alkanes/alkenes and molecular hydrogen were observed in the injection well up to month 7, indicating ZVI activity throughout that time period (Chu 2005).

GlaxoSmithKline, Research Triangle Park, NC: The site detailed in this particular case study was located in Research Triangle Park, NC, and owned by GlaxoSmithKline (GSK). While GSK was supervising the cleanup of the RCRA regulated site, contamination could be attributed to industrial activities carried out by the former property owners. A waste disposal area was previously located on site but was cleared out prior to the study by waste excavation. Despite the waste removal, residual contaminants remained.

While multiple chlorinated volatile organic (CVOCs) were detected at the site, TCE was the primary contaminant of concern. The groundwater concentration of CVOCs reached 14,000 ug/L in the injection well prior to the study. Concentrations in the immediate vicinity of the former waste disposal area most likely exceeded this value (Zhang 2003). The site was located in the Durham subbasin, an area composed of Triassic age sedimentary bedrock.

Groundwater was located approximately 30 feet bgs within the interbedded siltstone and sandstone. The injection site for the case study was located about 125 feet downgradient from the former waste disposal site mentioned previously. Two downgradient monitoring wells were installed approximately 22 and 43 feet from the injection site (Zhang 2003, Glazier 2003).

BNIPs and chemical oxidation were both considered for remediation; however BNIPs were ultimately chosen for the job. Part of the reasoning behind that decision was that the anaerobic bacteria, Dehalococcoides ethenogenes, had been detected at the site in early characterization. As mentioned earlier, the lowering of the groundwater ORP has been shown to stimulate the degradation of CVOC by bacteria (Glazier 2003).

The BNIPs used in the study consisted of iron with a palladium coating and were prepared by researchers at Lehigh University. A slurry was mixed on site with a BNIP concentration of 1.9 g/L. The total slurry volume was 6,056 L or 1,600 gallons, which translated to a total of 11.2 kg of iron to be injected. The injection was carried out over a span of three days at an injection rate of 0.6 gallons per minute (gpm) (Zhang 2003).

Following the injection period, the treatment accomplished over a 90% reduction in the total concentration of CVOCs in the injection well. This change occurred within several days of the treatment. A similar reduction in contamination occurred downgradient in the monitoring wells; however the decontamination process took longer (about 40 days to level out). Individual PCE, TCE, and DCE concentrations reached groundwater quality standards within six weeks of remediation and no significant amounts of hazardous daughter products were detected (Zhang 2003, Glazier 2003).

Naval Air Station, Jacksonville, Florida: The Naval Air Station (NAS) located in Jacksonville, Florida, has been operational since 1940. The site of interest for this case study, H1K, was positioned near the centre of the NAS. The source of contamination for the site was two underground storage tanks

(USTs), Tank A and Tank B. While they were excavated and capped in 1994 along with the associated pipelines, they were expected to be the source of contamination in the area. Prior to removal, two USTs stored waste solvents in addition to other substances received from a wash rack, manhole, and various other operations. Following removal, soil samples taken in 1995 confirmed that the removal of tanks and pipelines accomplished clean closure for unsaturated soil. The main groundwater contaminants detected, TCE and TCA, had accumulated around the former location of Tank A (Gavaskar 2005).

From 2000 to 2001 a chemical oxidation method was used to treat the source area. Rebound of the dissolved phase concentration occurred after each treatment. In 2002, the recharacterization of the site took place to reassess the level of contamination. Maximum soil concentrations of 25,300 ug/kg, 4,369 ug/kg, and 60,100 ug/kg were detected for TCA, PCE, and TCE, respectively. Groundwater maximums reached 173 ug/L, 5,520 ug/L, and 1,350 ug/L for PCE, TCE, and DCE respectively. The concentrations indicated the potential presence of DNAPL, which would lead to the rebound experienced (Gavaskar 2005).

Groundwater was present in a shallow aquifer located approximately 7 to 24 feet bgs.

Characterization of the unsaturated zone at the site indicated a fairly uniform composition of medium grained sand and sandy fill. Directly below the water table, between 6 and 12 feet bgs, a thin layer of clayey and/or silty sand rested on a fine to medium silty sand layer positioned 10 to 17 feet bgs. Between 20 to 24 feet bgs, the composition shifts to a larger amount of silt and clay, and then at 24 feet becomes mostly clay to a depth of 54 bgs (Gavaskar 2005). PARS Environmental's product, NanoFe Plus, was chosen for use.

The BNIPs consisted of 99.9% iron and 0.1% palladium by weight. The BNIPs were distributed throughout the site using both direct injection into known "hot spots" and closed loop recirculation. Direct-push technology (DPT) was used for the injections, where a suspension of BNIPs at 10 g/L was pumped directly into the DPT boor holes and allowed to discharge via gravity flow. Four injection and three extraction wells were used for the recirculation system. A BNIP concentration of 2 g/L was implemented initially in the recirculation process and then increased to 4.5 g/L (Gavaskar 2005).

Results of the remedial process varied widely from well to well. The recirculation process appeared to enhance desorption of contaminants into the dissolved phase. Many wells achieved over a 65% decrease in concentrations of parent VOCs within a short period of five weeks. All wells showed the present of "daughter" products, such as DCE, where in some cases the concentration of daughter products rose rapidly after injection and then decreased over time. An increase in ethane and ethane concentrations accompanied decreases in daughter products, indicating complete conversion (Gavaskar 2005).

Some source zone wells however, experienced a rise in both TCE and DCE concentrations after injection. This likely stems from poor distribution of the BNIP slurry and possible displacement of dissolved TCE. Of additional note, while the presence of ethane/ethane and C4-hydrocarbons, such as acetylene, indicate some abiotic degradation, the predominance of anaerobic reductive products such as 1,2-DCE suggests that much of the "parent" degradation occurred through microbial action. However, a well 20 feet downgradient of the source zone reported almost a 99% reduction in contamination, signifying that a significant amount of BNIPs potentially migrated out of the treatment zone (Gavaskar 2005).

Overall, the treatment provided some benefit to the site. In the end though, it was unclear how much of the decontamination that occurred could be attributed to abiotic degradation. nZVI and BNIPs have been identified as microbial enhancers and it is likely that a significant amount of the VOC destruction occurred through microbial pathways. It was speculated that some of the BNIPs might have become passivated before injection and also that the concentration of BNIPs in the slurry was not high enough (Gavaskar 2005).

Launch Complex 34 at Cape Canaveral Air Force Station, Florida: From 1960 to 1968, Launch Complex 34 (LC34) served as a launch site for Saturn Rockets. Ground contamination most likely occurred during that period and can be attributed to the rocket cleaning methods employed. Documented evidence indicates that rockets were cleaned with solvents including TCE directly on the launch pad.

At the time of cleaning, excess solvents were allowed to evaporate, penetrate the subsurface directly, or migrate to surface drainage pits as runoff material. Following the cessation of the rocket launching program at LC34 in 1968, the site was abandoned and became overrun with plant life. The contamination that occurred during the operational period resulted in the presence of DNAPL at the site prior to the study. TCE represented the main component of DNAPL contamination. However, cis-1,2,-dichloroethene (cDCE) and vinyl chloride, both products of the natural biodegradation process of TCE, were also detected in the groundwater. (Quinn 2005). The study itself was conducted on a small plot of land near the Engineering Support Building, a site of known DNAPL presence. The site contained a surficial aquifer and a semi-confined aquifer, where the groundwater TCE concentrations approached solubility (of TCE).

The surficial aquifer, covered by a clay unit, started at the water table, which began at a range of 3-7 feet bgs. Three layers comprised the surficial aquifer, an upper sand unit, a middle fine-grained unit, and a lower sand unit.

The upper sand unit extended from the ground surface to between 18 and 25 feet bgs and contained a mixture of medium to coarse-grained sand and crushed shells. At the bottom of the upper sand unit, the middle fine-grained unit began, extending to about 30 feet bgs and comprised of "gray, fine-grained silty/clayey sand," (Quinn 2005). The lower sand unit extended from the middle fine-grained unit to about 45 feet bgs. The injection of EZVI was planned for the upper sand unit (Quinn 2005).

Both soil and ground water samples were collected before the start of the injection to assess contamination levels. Six soil cores were obtained and samples from 21 monitoring wells were taken to provide pre-injection data. In an effort to maintain hydraulic control of the groundwater within the site area, a groundwater control system was installed. The system also served to assess the flux of DNAPL to groundwater during treatment. The EZVI, consisting of 44.3% water, 37.2% oil, 1.5% surfactant, and 17.0% nZVI by weight, was prepared on site. A total of about 670 gallons were injected across the entire pilot test area, which contained eight separate injection wells. The injection volumes across the eight wells were adjusted based on contamination concentration (Quinn 2005).

Based on the soil samples from the upper sand unit, significant reductions of TCE concentrations upwards of 80% were achieved at most of the soil boring locations. Problems in remediation were encountered at two wells, where it appeared that following injections, the EZVI migrated upwards, thus not reaching the contamination zone. It was suggested that the injection technique used, pressure pulse technology, contributed to the upward migration. Linear interpolation and kriging, a "statistical interpolation method for analyzing spatially variable data," also were used to estimate the overall success of DNAPL removal. Linear analysis estimated an initial total mass of 17.8 g of TCE with 3.8 g of DNAPL. This method then approximated an 85% reduction in TCE and TCE DNAPL mass. Kriging produced a value of 28 g of TCE for the initial mass and an average reduction of 58%. It was also observed that the total mass of TCE present below the upper sand unit, in the middle fine-grain unit, also decreased. An increase in "daughter" products however, points to possible biotic degradation in the lower soil unit (Quinn 2005).

Groundwater samples were also measured in the experiment and showed a marked reduction in contamination. A variance of contaminant destruction between 57 and 100% was observed in TCE concentrations between the various observation wells. Samples also displayed considerable increases in cDCE, VC, and ethane. The presence of cDCE and VC in the groundwater indicated the occurrence of biotic degradation. Combining the observation of daughter products in the soil and groundwater implied that biodegradation was responsible for a significant portion of TCE destruction (Quinn 2005). Overall the case study reported successful results for the remediation of TCE in both soil and groundwater. The study, however, reported inconclusive evidence on the actual mechanism of the TCE degradation and was unable to account for the amount of TCE degraded by the EZVI versus biotic degradation (Quinn 2005).

Toxicity and Safety Concerns

Generally, concerns regarding the toxicity of this technology have been mild. The confidence in safety is largely due to the fact that iron oxides formed during remediation are already present in the ground as rust, coupled with the fact that nano iron particles do not exhibit radically new properties. No studies reporting the safety and toxicity of iron nanoparticles or bimetallic particles have been published. However a study is due out within the next few months. The preliminary studies in daphnids found that nano-iron exhibited a similar toxicity level to that of the bulk form (Oberdorster conversation 2005). These findings are significant in that they support the previous theories of toxicity. In a recent review, Oberdorster et al. advised that studies not just concern toxicity as related to wildlife and human, but also focus on benthic and soil flora and fauna, "the basis of many food chains, which could be dramatically affected by nanoparticle injections." (Oberdorster 2005).

Bimetallic Particles and Other Metals

While scientists and contractors have developed and tested nZVI and palladium-coated nZVI particles on a large scale, many other metals and bimetallic combinations that can serve as substitutions exist. Metals such as zinc and tin possess similar reduction capabilities of iron (Boronina 1995). Like iron, these metals are converted to metal oxides in the decontamination process. Other metals have been combined with iron as well to produce similar results. Both iron-nickel and iron-copper bimetallic particles have been demonstrated to degrade trichloroethane and trichloroethene (Lien 2001, Schrick 2002). Another example is iron-platinum particles, which possess similar capabilities in degrading chlorinated benzenes (Lien 2001).

In producing bimetallic nanoparticles, inert base metals, such gold and aluminum, also have been used. Unlike iron, aluminum and gold particles are non-reactive or inert. Subsequently, they can only serve as base metals used in conjunction with a catalyst coating. Despite the lack of reactivity, base metals have been shown to increase the catalytic rates of the coating metal (Nutt 2005 and PARS 2005). In terms of metal catalysts able to be applied as coatings, many substitutions exist for palladium.

Platinum, silver, cobalt, copper, and nickel represent alternatives tested in the lab (Nutt 2005, Zhang 2003). While these catalysts possess the ability to work alone, coupling with other particles can enhance degradation and catalytic rates. It then follows that the various base metals can be fused with different noble metals to form a multitude of combinations. It should be distinguished that when metal catalysts are coupled with reactive metals such as iron, they serve to enhance the reactive properties of the iron. In contrast, when metal catalysts are coupled with non reactive metals such as gold, they act as the catalysts themselves.

According to the latest information, palladium-coated gold nanoparticles have emerged as the one of the most promising alternatives to nZVI and palladium-coated nZVI. As mentioned previously, palladium-coated gold nanoparticles differ from palladium-coated nZVI in that, palladium-gold particles act as catalysts instead of reactants.

The gold particles, which serve as the base metal, do not actually react with organic compounds, nor do they act as catalysts. Gold does, however, augment the catalytic ability of palladium. The enhancing ability of gold, while not yet clearly understood, has been proven by the fact that Pd-on-Au particles generate increased rate constants over 100 times greater than those for palladium alone (Nutt 2005).

In a recent study, Nutt et al. created Pd-on-Au nanoparticles and tested their ability to hydrodechlorinate trichloroethene in water (Nutt 2005). Because the Pd-Au particles catalyze the "hydro" dechlorination of TCE, hydrogen gas is necessary for the reaction to proceed. While scientists have studied gold and palladium particles as catalysts for a variety of other reactions, Nutt et al. appear to be the first in applying the catalytic abilities to remediation. Nutt et al. synthesized the catalysts by first creating gold nanoparticles. Using the citrate reduction method, scientists produced an aqueous fluid containing gold nanoparticle suspensions (Au Sol). To form the bimetallic particles, Nutt et al. added palladium salt in conjunction with a reducing agent to the Au sol (Nutt 2005).

The experiment compared the catalytic ability of palladium nanoparticles, gold nanoparticles, palladium-coated gold nanoparticles, palladium-alumina catalysts, and palladium-gold catalysts on alumina. Within the category of the Pd-on-Au particles, Nutt et al. also compared the effect of palladium coverage on activity. Results concluded that a sub-monolayer of palladium on gold produced the highest catalytic rates. An increase in Pd loading resulted in decreased rate constants, leading to the conclusion that once palladium coverage reached or surpassed one monolayer, the particles behaved as regular palladium particles (Nutt 2005).

Remedial Applications

At this point in time, the potential remedial applications of palladium-coated gold nanoparticles have been tested only in the lab for the capability to hydrodechlorinate trichloroethene and other chlorinated organic compounds in water. Due to the catalytic nature of the Pd-Au particles, they are not consumed in the reaction, and thus can be reused multiple times (Nutt 2005). This fact differentiates them from nZVI and Pd-nZVI particles, which are oxidized by contaminants such as COCs and subsequently passivated.

The ability to reuse Pd-Au particles multiple times, coupled with the product cost, greatly impacts the potential remedial applications of the particles. In a pound for pound comparison, nZVI particles cost much less than Pd-Au nanoparticles based on the price of the raw materials, but by capitalizing on the reusability of the Pd-Au, a smaller amount of particles can used in comparison to nZVI. Thus, in situ treatments involving slurry injections, where particles dissipate into the ground, work well for nZVI and BNIPs because particle recovery is not necessary. However, these applications would not be cost-effective for Pd-Au particles (Wong Conversation 2005).

Presently, development of remedial applications is focusing on two different innovations. The first technology involves mounting the nanoparticles onto membranes, allowing contaminated ground water or waste water to be pushed through the membranes for remediation. The second technology entails binding the nanoparticles to a powder.

This technique enables particle recovery from solution by filtration, a process unachievable with individual nanoparticles (Wong Conversation 2005). Field simulation studies designed to test the ability of Pd-Au particles to remediate groundwater contaminated with TCE are underway at Rice University (Wong Conversation 2005). Plans to expand the study to include other contaminants are also in development.

The list of potential contaminants includes other chlorinated compounds, fluorinated compounds, PCBs, nitrates, and potentially inorganic compounds such as arsenic (Wong Conversation 2005, Morello 2005).

Bibliography

Altmann, J. & Gubrud, M.A.: *Discovering the Nanoscale*, IOS Press, Amsterdam, 2004.

Andrew, N.: *Foundations of Nanomechanics: from Solid-state Theory to Device Applications*, Berlin, New York, 2003.

Attebery, B.: *Nanoculture: Implications of the New Technoscience*, Intellect Books, Bristol, 2004.

Bainbridge, W.S. : *Societal Implications of Nanoscience and Nanotechnology*, Dordrecht, Kluwer, 2001.

Bast, R.C. Jr.: *Cancer Medicine*, Decker, Hamilton, 2000.

Berube, D.M.: *Discovering the Nanoscale*, IOS Press, Amsterdam, 2004.

Brosnan, J.: *The Primal Scream: A History of Science Fiction Film*, Orbit Books, London, 1991.

Carroll, J.S.: *Social Science Research Methods for Assessing Societal Implications of Nanotechnology*, Dordrecht, Kluwer, 2001.

Caviedes, C.: *El Nino in History: Storming Through the Ages*, University Press of Florida, Florida, 2001.

Credi, A.: *Molecular Devices and Machines: a Journey into the Nano World*, Wiley-VCH, Weinheim, 2003.

Crow, M. & Sarewitz, D.: *Societal Implications of Nanoscience and Nanotechnology*, Dordrecht: Kluwer, 2001.

Daniel Ratner: *Nanotechnology: a Gentle Introduction to the Next Big Idea*, Upper Saddle River, Prentice Hall, 2003.

Dobzhanshy, T.: *Genetics of the Evolutionary Process*, Columbia University Press, New York, 1970.

Drexler, K. Eric: *Engines of Creation: The Coming Era of Nanotechnology*, Anchor Books, New York, 1986.

Dyson, Tim: *India's Historical Demography: Studies in Famine, Disease, and Society*, Riverdale, Westwood, MA, 1989.

Eamonn Gearon: *The Sahara: A Cultural History*, Signal Books, UK, 2011.

Frank J. Owens: *Introduction to Nanotechnology*, Hoboken, J. Wiley, 2003.

Glimell, H.: *Dynamics of the Emerging Field of Nanoscience*, Dordrecht, Kluwer, 2001.

Glimell, H.: *Societal Implications of Nanoscience and Nanotechnology*, Kluwer, Dordrecht, 2001.

Haber, L. F.: *The Chemical Industry: 1900-1930, International Growth and Technological Change*, Oxford: Claredon Press, 1971.

Hessenbruch, A.: *Nanotechnology and the Negotiation of Novelty*, Amsterdam, IOS Press, 2004.

James Lewis: *Prospects in Nanotechnology: Toward Molecular Manufacturing,* John Wiley & Sons, New York, 1995.

Karn, Barbara: *Nanotechnology and the Environment: Applications and Implications,* American Chemical Society, Washington D.C., 2004.

Keblinski, P. : *Encyclopedia of Nanoscience and Nanotechnology*, American Scientific, Los Angeles, Calif, USA, 2004.

Liu W.K., Karpov E.G., Park H.S.: *Nano Mechanics and Materials,* Wiley, Weinheim, 2006.

Mayr, E.: *Change of Genetic Environment and Evolution*, Allen and Unwin, London, 1954.

Mody, C.C.M.: *How Probe Microscopists Became Nanotechnologists, Discovering the Nanoscale*, IOS Press, Amsterdam, 2004.

Norio Taniguchi: *Nanotechnology: Integrated Processing Systems for Ultra-precision and Ultra-fine Products,* Oxford University Press, New York, 1996.

Orth, R. N.: *Nanometer Scale Antibody Patterning for Directed cell Immobilization and Stimulation,* Cornell University, Ithaca, 2002.

Pressman, J.: *Nano Narrative: A Parable from Electronic Literature*, Intellect Books, Bristol, 2004.

Pressman, J.: *Nanoculture: Implications of the New Technoscience*, Intellect Books, Bristol, UK, 2004.

Ratner, Daniel: *Nanotechnology: a Gentle Introduction to the Next Big Idea*, Prentice Hall, Upper Saddle River, 2003.

Robison, W.L.: *Discovering the Nanoscale*, IOS Press, Amsterdam, 2004.

Sarewitz, D.: *Nanotechnology and Societal Transformation*, Dordrecht, Kluwer, 2001.

Stephen M. Goodnick: *Transport in Nanostructures,* Cambridge University Press, Cambridge, 1997.

Tenner, E.: *Nanotechnology and Unintended Consequences*, , Dordrecht, Kluwer, 2001.

Vasudev Bhatnagar: *Cell Science and Technology*, Campus Books, Delhi, 2009.

Vermaas, P.E.: *Discovering the Nanoscale*, IOS Press, Amsterdam, 2004.

Weil, V.: *Ethical Issues in Nanotechnology*, Dordrecht, Kluwer, 2001.

Index